Handbook of Lean Manufacturing in the Food Industry

Michael Dudbridge

National Centre for Food Manufacturing
University of Lincoln
United Kingdom

T0305408

⊛WILEY-BLACKWELL

A John Wiley & Sons, Ltd., Publication

This edition first published 2011.
© 2011 by Blackwell Publishing Ltd.

Blackwell Publishing was acquired by John Wiley & Sons in February 2007. Blackwell's publishing program has been merged with Wiley's global Scientific, Technical and Medical business to form Wiley-Blackwell.

Registered Office
John Wiley & Sons Ltd, The Atrium, Southern Gate, Chichester, West Sussex, PO19 8SQ, UK

Editorial Office
9600 Garsington Road, Oxford, OX4 2DQ, UK
2121 State Avenue, Ames, Iowa 50014-8300, USA

For details of our global editorial offices, for customer services and for information about how to apply for permission to reuse the copyright material in this book please see our website at www.wiley.com/wiley-blackwell.

Library of Congress Cataloging-in-Publication Data

Dudbridge, Michael.
  Handbook of lean manufacturing in the food industry / Michael Dudbridge.
     p.   cm.
  Includes bibliographical references and index.
  ISBN 978-1-4051-8367-3 (pbk.)
1. Food industry and trade–Management–Handbooks, manuals, etc.   2. Food industry and trade–Waste minimization–Handbooks, manuals, etc.   3. Lean manufacturing–Handbooks, manuals, etc.   I. Title.
  TP373.D83 2011
  664.0068′5–dc22

                                                       2010041176

A catalogue record for this book is available from the British Library.

This book is published in the following electronic formats: ePDF 9781444393101; Wiley Online Library 9781444393125; ePub 9781444393118

Set in 10/12.5pt Times by SPi Publisher Services, Pondicherry, India

1   2011

Contents

Introduction

How to use this handbook

Welcome to the *Handbook for Lean Manufacturing in the Food Industry*.

This book has been written to act as a source of information and ideas for people working in the demanding world of food manufacture. The aim is to provide food industry personnel with methods of increasing efficiency, reducing waste, lowering costs and improving control in their factories. It will also point the way to less breakdowns, reduced quality faults, increased teamwork and improved profits.

This book has been written in a style that makes it easy to read and contains practical examples of the issues addressed. Illustrations, photos and tables are used to aid understanding of the topics discussed and the book is broken into bite-sized sections that can be digested in just a few minutes each.

This book has been designed with two methods of use in mind. First, it could be read from cover to cover, which would be a great way to get the full picture and understand all areas and techniques. The second method of use is to dip into the sections that address issues you are encountering at the moment in your factory. Dipping in is made easy by the comprehensive indexing system and the fact that there are referrals from one section to another within the handbook.

For example, a section on waste reduction may also refer you to a section on machinery breakdowns. These links are not compulsory; if you think you have enough information to make a difference to the business, your time would probably be best spent on the shop-floor talking about trying out the ideas together.

> It is recommended that you read this book with a pen and notepad to hand for you to jot down ideas as you read. This is a great method to make sure that you do not forget that great idea! Look out for the question boxes. By answering the questions posed, it might help you recognise what is required to improve performance.

You can give yourself a reminder to "see Mary tomorrow about the raw material quality problem" or "start to measure downtime on the packing machine to find out the top three reasons it keeps stopping."

You should use the book as *your* record of *your* issues, with their possible solutions. If you make this book your own, it will serve you well for years, as you and your company continue to improve and you move on to the next issue, and then the next one.

The National Centre for Food Manufacturing

In 2007, it was decided by the University of Lincoln in the UK to build a training and development centre at Holbeach in Lincolnshire. The centre was to provide a centre of excellence in the training and development of people employed in the food manufacturing industry. By entering into partnerships with the Processing and Packaging Machinery Association (PPMA) and major manufacturers of processing and packaging machinery, the university created a centre that was at the front of processing and packaging technology.

In March 2009, the centre was opened and now delivers training and education to food factory personnel from throughout the UK. The centre is used by people employed in the business of food manufacture, to increase their knowledge and skill in all aspects of the manufacture of food on a large scale.

The centre continues to develop and is forming partnerships with more equipment and service providers, to make sure that it contains the very best of technology in a modern food factory environment (see Figure a.2).

Figure a.2   The NCFM at Holbeach boasts state-of-the-art food manufacturing technology in a modern food factory environment.

About the author

Michael Dudbridge is Principal Lecturer in Food Manufacture at the National Centre for Food Manufacturing, Holbeach Campus, University of Lincoln, UK. After graduating in Food Technology from Reading University in 1979, Mike has worked in and around the food industry for most of his working life. He has a very understanding wife, Linda, and two great kids, Ben and Samantha, both of whom are destined for careers in engineering.

Mike joined the university in 2006 as a lecturer in food manufacturing and is based at the specialist Food Campus at Holbeach in Lincolnshire, UK.

Prior to joining the university, Mike worked in the food industry in various managerial roles for major food manufacturers in the UK. Most roles focused on operations and manufacturing, but occasionally in technical areas too.

Recently Mike's role grew to include responsibility for commercial partnerships and industrial projects and also the management of the new National Centre for Food Manufacturing at the Holbeach Campus, a state-of-the-art training and research facility for the food and drink industry: http://www.lincoln.ac.uk/holbeach

Mike is a specialist in Lean Manufacturing techniques and automation and considers himself a production manager who teaches rather than a teacher of production management (see Figure a.3).

Acknowledgements

A handbook of this type is created following a lot of help, assistance and support from the people I have worked with throughout my career in food manufacturing. For all of their support I would like to thank my colleagues and friends – without them I would not have had the experiences required to complete this work. I would like to thank the publisher for all their understanding, guidance and support.

I would especially like to thank my understanding wife and family who allowed me to take my laptop on holiday and excuse the fact that writing a book is a time consuming affair.

Finally I would like to thank Rhian Burge of Ishida Europe, the staff of the National Centre for Food Manufacturing and the staff of Bakkavor for providing me access to many of the photographs contained in the handbook.

Acknowledgements

A handbook of this type is created following a lot of help, assistance and support from the people I have worked with throughout my career in food manufacturing. For all of their support I would like to thank my colleagues and friends — with out them I would not have had the experience required to complete this work. I would like to thank the publisher for all their understanding, guidance and support.

I would especially like to thank my understanding wife and family who allowed me to take my laptop on holiday and use of the fact that writing a book is a time consuming effort.

Finally I would like to thank Khan Bingo of Iskada tongue, the staff of the National Centre for Food Manufacturing, and the staff of Bakkavor for providing me access to many of the photographs contained in the samples.

Chapter 1

The food industry

This chapter looks at the food manufacturing industry and begins by looking at some of the reasons behind Lean Manufacturing.

Pressures in the food manufacturing industry

As you begin to read this book, it is important that you spend a few minutes putting your work into the bigger picture, so that you can appreciate the reasons behind some of the issues you have now or could have in the future. This will also help you come up with better solutions that will sustain you for longer.

Reasons why the food manufacturing industry is unique

Political food

The food industry is very important to a lot of people in this country. The supply of safe, affordable and plentiful food is essential to the well-being of the nation, as well as to the political success of the people running the country.

There is hardly a week goes by without a food story hitting the headlines. This could be a food safety scare, an increase in prices, food retailers making massive profits or even a secret camera being taken inside a food factory. Add to that the public's concerns over packaging and waste, food miles and obesity, and you can see just how big a part of people's everyday life is impacted by the food industry.

Governments know that people are interested in the safety and price of the food they eat. Everyone also appears to be increasingly interested in where their food comes from, how it is reared or grown and if it has the required nutritional content. Because of this need to deliver food safety and value, the food industry is heavily regulated to ensure that nothing goes wrong.

The food business

The food industry is a massive business. In order to keep the nation fed, a huge quantity of food has to be shipped each day from the farms to factories and then to the retailers and ultimately the consumers. The need for reliability in this supply chain is paramount for

Handbook of Lean Manufacturing in the Food Industry, First Edition. Michael Dudbridge.
© 2011 Blackwell Publishing Ltd. Published 2011 by Blackwell Publishing Ltd.

Figure 1.1   Food factories can be highly complex, high performing and deliver what the customer wants when the customer wants it.

the consumer to be satisfied at a price they are willing to pay. Food retailers know that their shoppers are very sensitive to issues in the supply chain. If a food product is unavailable in a store, the consumer may well look elsewhere and the store will have lost business, not only for that product but for the rest of the shopping basket as well. Availability of products in store is therefore a very important factor in food factories; ensuring that deliveries are made "on time" and "in full" is a major task.

The food retailers also compete in terms of the prices of the products on sale in their stores. This pressure on prices is passed back to the food factories in the form of small profit margins. Low margins and high delivery performance will place stress and strain on food businesses and the people that work in them. In order to perform well in the food business, costs have to be controlled and minimised, and performance has to be at a high level and consistent. A food factory that under-performs in any of the areas of its activity will soon run into difficulties with its customers, as in any other businesses. The difficulty in the food manufacturing industry is that the level of high performance has to be maintained day in, day out, with low profit margins meaning that slacking and duplication in the systems cannot be afforded. A food factory cannot afford to have a spare production line, "just in case of a breakdown". It cannot afford to have spare raw material stocks, "just in case we have a process fault and have to reject some products". It cannot usually afford to have spare anything! The work of a food factory has to be right first time, every time (see Figure 1.1).

Food fashion

Consumers of food – that is everyone, are constantly being tempted to buy new and exciting foods. There are launches of new food products on an almost daily basis, from companies trying to persuade consumers to buy their products. A large number of food products do not survive for long in the marketplace, as they are constantly replaced by a "new improved" version, or the sales decline because the consumers have moved on to a new product. For the food factories, this constant stream of new products has its own set of demands and pressures that need to be controlled. The food retailers are constantly trying to excite their customers with new products and an every increasing range of foods. The massive variety of foods available in retail outlets has a big impact on the work of food manufacturers. The number of different raw materials and processes that are needed to manufacture the products means that food factories are becoming increasingly complex places to work. The warehouses for raw materials and packaging need to be well managed, to maintain control of the quantity and location of the stock. There is nothing more frustrating in a food factory than the words "I know we have got some ... now where did I see it?"

The complexity of the factory will also have an impact on the methods of production. The rapidly changing fashions in food mean that no one in the food supply chains want to get caught with excess stock. There is nothing worse than having a large stock of a specialist raw material when the fashion changes, except perhaps having a large stock of packaging material! As a result, manufacturing operations have to be set up to be very flexible to meet demand. The way that food manufacturers have responded to the food fashion aspect of the business is to make little and often. This has always been a feature in short shelf-life foods, such as chilled products, bread and fresh produce, but is increasingly used in factories that manufacture long-life products, such as frozen and canned foods, biscuits and preserves.

Food fashion has taken food manufacturers into the area of low stocks and flexible manufacturing. This appears on the shop-floor as short production runs, multiple line changeovers, complex material controls and a need for precise, right-first-time production (see Figure 1.2).

Why do we bother?

"Why don't we make something else for a living?"

The food industry has great rewards, if you get it right. The food industry is a massive business and if a factory performs well it will grow. The food industry is seen as "recession proof"; people have got to eat! Indeed, during periods of economic downturn, the food that people eat has been know to increase in value, as they look for comfort and shelter from the economic storm.

Questions

Have a think about your factory – answer the following questions:

- How many new products have you launched in the last year?
- When was the last time your factory failed to deliver what your customer wanted?
- When was the last time you saw a food story on the TV?

Figure 1.2   The food fashion, massive variety in retail gives complexity throughout the supply chain. It is one of the challenges of a food factory to operate and thrive in this complexity, while still remaining profitable. Companies have a variety of strategies for meeting these challenges. Lean Manufacturing is one of those techniques that helps.

The major retailers can be very demanding of their suppliers in terms of technical standards, delivery performance and price, but they do give a manufacturer access to a massive customer base. The volume of product sold through the major retailers allows the manufacturer to run a large-scale operation and as a result they benefit from lower costs. The economies of scale in the food business mean that the more you manufacture, the cheaper each item becomes. The price, to the manufacturer, of raw materials and packaging materials will reduce as the volume increases. The costs of energy and distribution are also related to the quantity bought, therefore, the bigger the business, the lower the prices.

Well, at least that is the theory. The food business is very complex so sometimes the normal rules do not apply for short periods, but in the medium and long term the rules will have to be applied if businesses are to make a reasonable profit and thrive.

There follows some more reasons why the food industry is unique.

Food materials

The raw materials used in the manufacture of food are natural products that are either grown or reared on a farm. Occasionally, manufactured or artificial ingredients are used, but this is becoming increasingly rare as the industry caters for consumers that want their food made

Figure 1.3   Food as a raw material is one of the factors that make the application of Lean Manufacturing different in the food industry to any other. The variety and inconsistency of the materials arriving at a factory, coupled with the fashion food and political parts of the picture, make food a fascinating example of manufacturing success (picture courtesy of Ishida Europe).

without these ingredients and so regulations control their use. If the material is meat for a sausage or flour for a loaf of bread, sugar for a biscuit or vegetable oil for pastry, the vast majority of raw materials originate on the farm. These natural materials are not often uniform, so variation will occur from crop to crop or animal to animal. The variation will also occur from season to season and year to year as the growing conditions change due to rainfall, temperatures or if the practices on the farm alter. The food industry tries to produce products to meet the specification, but variation in raw materials means that the processes used will rarely be static and often need to be altered to compensate for the raw material variation. That said, a focus on your raw material suppliers will allow you to minimise the variation that hits your process and at the very least, with careful management, predict the effect of that next batch of flour or consignment of strawberries.

What all this means to a food factory, and those trying to manage the process, is that even if your process today is identical to your process yesterday, it does not necessarily follow that the same quality, yield, efficiency and waste levels will be achieved. For this reason, one aspect of Lean Manufacturing focuses on monitoring and control of process performance, so that corrective actions can be taken before a raw material variation becomes a rejected product (see Figure 1.3).

For all the above reasons, the use of Lean Manufacturing techniques in a food factory is different to those same techniques used in a car plant or a factory making mobile phones!

24/7

Pressure on profit margins and the need to supply large quantities of food has lead to the growth of increasingly large food factories, which are often running for 24 hours a day, 7 days a week. This high intensity of manufacture often means that those managing the shop-floor are part of a shift system and that their performance is reliant on the performance of others who they rarely meet. The need for teamwork in the food industry is vital to its success and that teamwork has to extend to people who are in a different department, on a different shift or even on a different site. Add to this the demands posed by the need for high performance and it can be seen that managing a food production operation is a highly demanding role that needs to be done well if performance is to be maintained, and is even more demanding if performance is to be improved. Later in this book, costs of manufacture are discussed in more detail. It is very important that all members of a team in such a demanding industry have knowledge of the pressures on the business, their role in the team and that the overall aims and objectives are clear. In a factory that is implementing Lean Manufacturing techniques, everyone must be given the opportunity to contribute to the improvement of the business. This is particularly the case in the food manufacturing industry, where high performance and low cost is not just demanded by the retailers, but is the only way that a business will survive and grow in such a competitive market.

Alternative methods of lowering costs

First, you should spend some time looking at the way that a food manufacturing business works from a financial point of view. An understanding of the costs of the business, and how they are monitored and controlled, will help in the application and understanding of Lean Manufacturing techniques.

In a food business, costs are the one factor that needs to be considered alongside each and every decision. Food safety, quality and service level are also vital in the day-to-day running of the business, but unless the costs are under control the business will have a big problem.

Keeping costs at, or better than, the predicted levels will allow the business to make good decisions and remain in control of all aspects of that business. The business will remain balanced, with decisions made for the short, medium and longer term each having their place. Businesses that lose control of costs can end up making decisions based just on short-term survival, with little or no spare resources to be able to plan for the future or ensure long-term performance.

It is when the costs of manufacture become inconsistent, or rise above what was predicted, that the issues start. One of the main areas of Lean Manufacturing is in cost control, and its application in the food industry is very important because of the tight profit margins. Corrective actions need to be taken very quickly to get costs back under control.

Before costs can be controlled and reduced, the different types of costs in your business and the way in which those costs behave must be considered.

The theory of costs

Costs have been traditionally divided into three types for the purpose of analysis. This is a simple way of gaining an understanding of costs in a manufacturing business and will allow you to better understand what is happening in your factory. You will know from your own experience that nothing in business is simple, so you need to start with a simple look at costs and how they work.

Variable costs

Variable costs are normally associated with production labour, raw materials and packaging materials. The money paid out by the business goes up as more product is made and goes down if the orders are low and the factory is less busy. The costs *vary* with the activity level of the business.

For example, in a bakery the more bread that is made the more flour is required. Thus, the size of the invoice from the flourmill will rise as more bread is made and fall when the factory is less busy. The same will go for packaging costs. The money paid out by your business varies with the activity level of the factory.

Fixed costs

Fixed costs do not change with the activity level (over a normal range of trade) and are generally associated with factory services, management and so-called overhead costs. The monthly rent paid for the forklift trucks does not change if the sales are down a bit or up a bit, though the factory may consider renting an extra truck to help cope with a big volume increase associated with a special promotion. In the normal range of trade, the cost of the trucks is said to be "fixed". The salary of the human resources manager is the same each month, even if the factory volumes are down by 10%. There are many fixed costs associated with running a manufacturing business, but as was said earlier in this chapter, the retailers and the consumers demand flexibility in terms of product type and also product volumes. The last thing a food manufacturing company needs is for the fixed overhead costs to be too big a proportion of the overall costs of the business. Fixed costs need to be minimised to ensure that the company can survive if sales volumes drop. The combination of high fixed costs and low profit margins will result in the business finding it difficult to survive a period of low sales.

Semi-variable costs

In practice, most costs are semi-variable. This means the costs do change as activity level changes, but that change is not even.

For example, labour cost will increase as the factory gets busier but there will come a point where people will be paid enhanced pay rates to encourage them to work extra shifts or work night-shifts or maybe weekends. Some factories have an agreement with their workers that overtime rates are not paid and the worker gets the same rate for all hours worked; this is an attempt by the factory to keep labour costs as variable as possible and to prevent costs escalating when the factory is busy. The semi-variable nature of labour costs is also shown when the factory is quieter. As business slowly falls, worker's hours

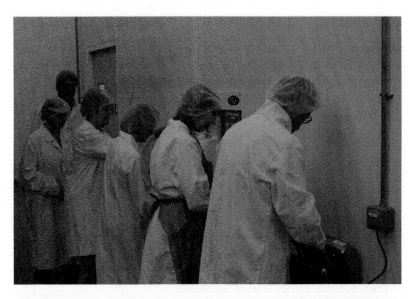

Figure 1.4   End of shift – clocking out – what was the cost of the hours worked during the shift. Was that cost variable or was part of the payroll semi-variable or even fixed?

will start to fall but there is nearly always a tendency for new tasks to be found to keep people busy until the end of their shift. If production is taking six hours instead of eight, there is always a temptation for one of two things to happen:

1. The six hours of work is stretched to make it last eight hours. This is often not a deliberate action but is the result of a lack of urgency in overcoming small hitches in the process.
2. The required work is finished in six hours but additional work is created to build up finished product or work in progress stocks. As stated earlier, increasing stock levels is not always a good idea in a food factory (see Figure 1.4)!

Another semi-variable cost is that of materials, food and packaging. When volumes increase over an extended period, it is possible for the company to negotiate with suppliers and to gain volume related discounts. For example, carrots are £300 per tonne, but if you take 20 tonnes, the price drops to £250 per tonne. The semi-variable nature of the costs of materials is also shown when the full cost of obtaining them is considered. For a food factory to buy materials, it has to place an order. The cost of raising the order (the time and effort of the buyer) is the same if the order is for 1 kilo or for 1000 kilos. The same applies to delivery costs. One pallet on a lorry is going to cost roughly the same as 26 pallets to deliver.

The final example of a semi-variable cost is one where the cost for something is made up of a fixed element and a variable element. For example, the costs for telephone services will often have a fixed equipment charge or line rental and then a variable cost of the calls made. The costs of compressed air in you factory will follow a similar pattern; there are fixed costs associated with the provision and maintenance of the compressor and pipelines, but a variable cost for the electricity to run the compressor. The same goes for refrigeration (see Figure 1.5).

Figure 1.5 Refrigeration compressors – a semi-variable cost; you can choose to switch them off to save electricity costs but there is a fixed cost element to keep them maintained, which is not related to the volumes of product made in your factory.

Now that you know a little more about costs and the way they work, look at some of the ways in which costs can be reduced. Try looking at some typical costs for a food factory; to understand what might be required to reduce them.

First, the variable costs:

Questions (see fig 1.6)

Think about the variable costs in your department or factory:

- Do they match the ones in the chart?
- How about the cost of each item on the list?
- The chart contains four questions marks – what costs would these be in your area?
- See if you can think of four more costs of production that vary with the amount of product made.

You can see from the pie chart (see Figure 1.6) that the variable costs in a food factory are centred on the materials and labour. The efficient use of these resources will reduce costs to a minimum level, but before considering the use, think about the prices paid for these things.

Food materials

The price paid for food materials will vary, depending on the material and the negotiating skills of the person doing the buying. There are two ways of reducing the price of food materials, you could choose to buy materials of a different specification or you can negotiate a better price for the same material.

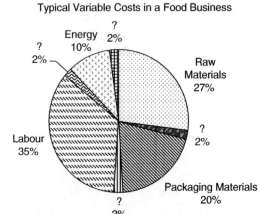

Figure 1.6   Examples of food manufacturing variable costs.

Different specification

Could be that the material you are currently using is too good (and expensive) for the job you are asking it to do. Would a different and cheaper cheese work just as well on our pizza? If the onions were changed from their current type to cheaper ones, would that make a difference to the quality of the ready meal? A great deal of care has to be taken with this kind of decision. The consumers of your products are a very valuable part of your business. Changing a material could have a detrimental impact on the product and sales volumes could reduce as a consequence.

Another aspect of working with a different specification is to change the recipe of your product by reducing the quantity of the expensive raw materials. Again this process needs to be well controlled by the business. Not only are there customers' views to consider, but there are also legal aspect to the contents of the food. Packaging will contain ingredients listings and nutritional values that must be met.

The final impact of changing material specification is on the process of making and packing the food. For example, it may have been decided to reduce the thickness of a packaging film to reduce its costs. A good idea, except when the new packaging was used in the factory for the first time it was apparent that the performance of the wrapping machine dropped. There were more rejected packs and more line stoppages. The cost saving on the packaging material was negated with increased cost in waste and poor productivity. This is an aspect of cost saving that will be considered again later.

Reducing costs is a bit like trying to put a tonne of marshmallows into a pillowcase. You squeeze in one part and it pops out somewhere else!

Questions

- Is there an opportunity in your business to use a different specification of materials?
- Would changing have any negative effect on performance?

Negotiate better prices

It is possible to negotiate better prices from your suppliers and reduce the costs to your business as a result. In the negotiation there may be a need to increase the quantity of product purchased from that supplier to secure the better deal. "If I was to buy all of my cheese from you, what is the best price you could give me?"

Better prices can also be negotiated by helping your supplier to reduce their costs. "If I buy my vegetable oil in 1 tonne tanks instead of 25 litre drums, you can reduce the price by 5%? If I take one delivery per week instead of three per week, you will reduce price by 9%?"

There are many mechanisms in place to reduce price of food materials; this is a specialist area of the food business, but people from the manufacturing areas can often suggest improvements that allow the negotiations to start. The trick here is to make sure that the reduced price does not cause increased cost elsewhere in the business. For example, buying very large quantities of materials will reduce the price of the materials (unless you try to buy so much it causes a shortage and the price rises)! But once you have purchased the material, you will have to store it and that will add to your costs in the warehouse. There is also the risk in the food industry of falling victim to food fashion and being landed with a stock of raw material for a product that is no longer wanted by the consumer. The final risk to the business is that in paying for the materials the business becomes short of money. The company has got loads of raw materials but no cash in the bank to pay the electricity bill! Both large and small companies have to keep a very close eye on the use of cash in their business, to ensure that the business remains solvent and able to meet its obligations.

Questions

It is a well-known fact that production managers always think that they could buy the materials used in their factories cheaper that the buyer! It is also well-known that buyers think they could run the factory better than the production managers!

- Is there any opportunity in your factory for the price paid for materials to be reduced?
- Make a note of three materials it would be worth investigating.

Packaging materials

The prices paid for packaging materials are often based on the complexity of the design and the weight of material in the packaging. The techniques used for food materials can also be applied to packaging materials, but in a slightly different way.

Questions

- When did your factory last have a good look at the packaging used on your products?
- Is your packaging dictated by your customers?
- Is there any scope in your business to reduce cost here?

Packaging redesign

By looking at the packaging format used for the food product, it is often possible to change the shape, size or design to reduce the cost. This is a method that is currently applied across a wide range of food sectors, as consumers want to see less packaging around their food products. The technique of light-weighting is one where the mass of packaging materials is reduced by making bags smaller, making trays thinner or reducing the gauge of cardboard. Light-weighting also looks to reduce the number of layers of packaging used for some products. A bag inside a box, then over-wrapped, is a great opportunity for packing cost reduction.

Printed packaging materials are expensive to manufacture. Simplification of the print design will reduce the price of the packaging. For example, moving from a full colour photographic image of the product to a simple three colour graphic design will reduce the price of the pack.

In some ways packaging has become more complex recently with the advent of Shelf Ready Packaging (SRP), sometimes called Retail Ready Packaging (RRP). A simple box has been replaced with one that has a more complex structure and often has quality graphics printed on it. These cases are often more expensive than the plain cartons they replaced. They also cause extra complexity when they are specific for a particular product and therefore cause a factory to have four or five stock items instead of one plain box. "Why were these SRPs introduced?" The answer is a piece of "whole supply chain thinking". The additional costs at manufacture are more than compensated for at the retail outlet by the system of "one-touch replenishment". Staff and efficiency savings in store save more than the additional cost of the SRP.

SRP can add cost and complexity to a manufacturing operation and save cost in the retail outlet. The retailer may be willing to pay more for a product in SRP.

Negotiate better packaging prices

Packaging, especially printed packaging, is very dependant on volume. The price per pack will tumble as volume increases, but there are risks associated with placing orders for large quantities of printed packaging simply to get the lower price.

There is a risk that the food fashion will change and the product stops selling or has to be changed. As a result, the printed packaging become redundant and has to be disposed of. The second risk is that a large packaging stock will have to be stored, looked after and kept in good condition prior to its use. There is a risk that the packaging material may deteriorate in store and not perform correctly when brought into the factory for use. The final risk is a financial one; the packaging material in your warehouse will have been paid for and that money is then tied up in large stock levels when it could be invested elsewhere in the business for a better return.

This chapter will now look at one of the major costs in the manufacture of foods that is within the control of the business and the production team.

The price of labour

This vital ingredient in the manufacture of food products is often missed when cost reductions are required. The price paid for labour is a large proportion of the variable costs for many food businesses. The usual way to reduce labour costs is to make the labour more efficient, so more is achieved with less labour. Automation of processes, improved planning and efficiency

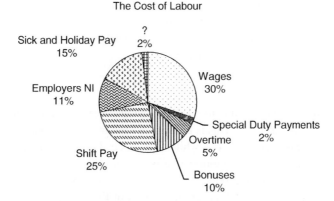

The Cost of Labour

Sick and Holiday Pay 15%

? 2%

Wages 30%

Employers NI 11%

Special Duty Payments 2%

Overtime 5%

Shift Pay 25%

Bonuses 10%

Figure 1.7 How the total cost of labour is made up.

improvements are a way of reducing the overall costs of labour and these will all be covered in greater depth later. However, let's consider the *price* of labour for the moment.

The total price of an hour of labour is made up of:

The rate of pay

This is the rate per hour worked. This can sometimes be variable for an individual, such as they get extra money for working unsocial hours, for overtime, or special responsibility pay for a certain task such as forklift truck driving or first-aid responsibilities. The reasons for increasing a pay rate are many and varied and have often been the subject of negotiations over many years.

Once the rate of pay has been decided, the factory price is not complete, because there are extra costs to be paid to arrive at the final price per hour.

Employer costs

These costs include items such as National Insurance payments, holiday pay and sick pay provisions. All of these costs add to the true price of labour. However, there are still other costs that need to be considered.

How about the cost of overalls and safety shoes and other personal protective equipment (PPE), the costs of subsidised canteens and of providing a secure well-lit car park?

Therefore, the price of labour in your factory is not simply the wage bill but includes a large range of costs that can often add up to a substantial sum. This makes the sections in this book that look at labour efficiency even more important to the profitability of your factory (see Figure 1.7).

You can see from the pie chart that the total cost of labour in a factory is made up of many parts.

Questions

- How do these categories compare with the price of labour at your factory?
- Are there other costs of employing people?
- Do the percentages shown reflect the picture at your business?

How do companies react to the price of labour? Indeed, is it ever possible for a company to reduce these costs? There are several options open to companies to try and reduce the price of labour within their business. Each method is fairly difficult to carry out and often meets with resistance of one kind or another.

Off shoring

Perhaps the most radical method is to relocate manufacturing operations to a part of the world with a lower labour price. This is unusual in food businesses, but examples of this technique hit the press every once in a while. The reduced price of labour in a new location has to be balanced with any additional costs associated with transport and storage of the raw materials and the final products. Off shoring in the food industry has been most popular with the multi-national food groups that have the ability and the resources to manufacture products almost anywhere in the world. Often the activity in the multi-nationals is centred on filling up the most efficient plants rather than just chasing the cheapest labour price.

Renegotiation

Discussions take place to try and reduce the price by negotiation. This could involve a reduction in shift allowances, overtime premiums or other supplements and would normally be bundled into a package with some wins and some losses for the individuals involved. Various techniques have emerged over the years to renegotiate the price of labour. Annualised hours have been introduced in some companies where the worker is contracted for a number of hours per year, with an element of flexibility in the hours per week worked. In return for the annualised contract, the worker receives a regular monthly salary that does not vary. This technique gives the worker a steady and predictable income in return for flexibility and the loss of premium payments for overtime. During recessions in the economy, when workers are in fear of losing their jobs, it has been possible to negotiate a reduction in the price of labour in some businesses. This is a difficult decision for all concerned, but is often part of a package of changes aimed at saving as many jobs as possible as a factory struggles to survive.

Pension

Some companies have looked hard at the costs of their pension schemes and have either shut them down or stopped more people from joining. The life expectancy of people is increasing and as a result the future liabilities of the pension funds have increased also. Companies have done the calculations and have often decided that the cost to the company cannot be sustained and they reluctantly change the pension arrangements for their staff.

National Insurance

There is not much a company can do to avoid this cost, but some have been eligible for help in the form of subsidies and payments from governments. Employer's National Insurance holidays are often used by government and regional agencies to encourage companies to set up or expand operations in an area that has high levels of unemployment.

The use of agency or contract labour

Agency labour is a big feature in many food companies and can often cost less per hour than for an employed person. Agency labour is where a worker is employed by an employment agency and is sent to your factory to carry out the work required on a temporary/casual basis. Agency workers are often employed in low skill jobs in the factory but can also be used as office temps in the administration of the business. The company pays a price per hour but has no responsibility for any other costs (with maybe the exception of PPE). The use of agency labour will be discussed later in this book, but for this section it is enough to know that agency labour is sometimes a useful tool in trying to reduce the price of labour in a business. The costs associated with the recruitment and induction of a new employee are eliminated.

Because of the low skill and low wage nature of a lot of food factory jobs, there is sometimes an issue with high staff turnover rates. Staff turnover is a measure of the length of time a new worker can be expected to stay with a company. The food industry has, traditionally, suffered very high staff turnover and the use of agency or contract labour can reduce the costs incurred in recruiting a person who stays for only a short period of time.

Modern agreements with agency worker providers can mean that the temporary worker arrives having had a full company induction, literacy and innumeracy tests, a medical, hygiene training and sometimes even the required PPE. The agency and not the food manufacturer incur all of that cost. The hourly price paid for the agency worker can, on the face of it, appear quite high but, as will be shown later, can still be a great advantage to a food factory.

Let's look at the fixed costs

There is no such thing as a fixed cost – some are just a bit more difficult to reduce than others.

This is a real quote from a wise old boss some years ago. He was, of course, perfectly correct; fixed costs are only fixed because no way has yet been found to make them variable.

It is the aim of most food businesses to reduce the amount of their fixed costs and make as much of their costs as they can into variable costs. In that way, the cost goes up when the business is busier and down when the business is quiet.

Remember, the fixed costs have to be paid no matter how busy or not the factory is, and over a normal range of sales volumes the cost stays constant. It is important to know about fixed costs in the application of Lean Manufacturing to the food industry. Companies have grappled with these issues for a long time.

Here are a few ways that food companies have handled the issue of fixed costs:

Fixed cost dilution

This is an effective method that is widely used. The fixed costs are not reduced but the size of the business is increased without the fixed costs increasing. The same fixed cost is diluted over a bigger business and so have a reduced impact on the profit margin. There are several ways in which companies have carried out this fixed cost dilution. They can ensure that the factory is manufacturing to full, or near full, capacity for the entire time. For example, the use of price promotions to increase the volume of product made will make the factory busier and this will help dilute the fixed costs over that bigger business.

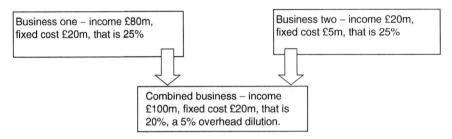

Figure 1.8   Fixed cost dilution through the acquisition of a second business.

A second method of diluting fixed costs is to spread the costs over several businesses. For example, the technical manager now looks after two factories instead of one. The fixed cost of the manager is spread over a bigger business. The final method of fixed cost dilution is sometimes called consolidation or rationalisation, This is a big move for any company and has massive consequences for all concerned. Instead of running two factories with two sets of fixed costs, one of the factories is closed and the work of that factory is transferred to the other. This increases the size of the business in the surviving factory and as a result the fixed costs of that site will be diluted over that bigger business. Examples of this are reasonably common in the food industry, among bigger companies; some examples have been seen of a company buying out one of its competitors to "get hold of the volume", and then to move that volume and close down the factory that has just been purchased.

For example, a business with a fixed cost of £20m per year and an income of £80m has a fixed cost of 25%. The business grows by the acquisition of a competitor that had a turnover of £20 m and fixed costs of £5m. The new bigger business was restructured in the first few months and the fixed costs returned to £20m for the new larger company. The fixed costs are now 20%. That is a 5% fixed cost reduction that is sometimes called dilution of over-heads. The £5m that has been saved in the new business is additional profit and is often used by acquisitive companies to finance the bank loan that paid for the acquisition in the first place. The structure of the food business is such that this kind of acquisition takes place often. The acquiring company sees an opportunity for rapid growth by using techniques to reduce fixed costs in the target business. A company with high levels of fixed cost can become a takeover target if the economic conditions allow (see Figure 1.8).

Fixed cost to variable cost

This is a group of techniques to reduce the fixed costs in a business and to make the cost variable, so that the cost can be controlled depending on the activity level of the factory. There are several ways in which this can be done. Rather than owning a resource that the business needs, it is rented on a short-term agreement or the service is bought in from a supplier. Of several examples, the largest one is probably transport. A few years ago, companies would run their own fleet of delivery vehicles to get their product to their customers. Some still do, but a large number of companies have decided that the fixed costs that are tied up in transport are large and difficult to reduce if the business is a bit slack. The vehicle still has to be paid for, even if it is parked in the yard. This fixed cost can be made a variable one by taking on a transport company to deliver the food. When the vehicle is needed it is paid for, when the vehicle is not needed it is not paid for; this is perfect variable cost control!

This technique has also been applied in other areas of fixed cost. Computer and telephone systems are often leased based on a usage basis, you pay when you use, you do not pay when the resource is not used. There are also examples of food production machinery rented under short-term agreements, maybe to provide extra capacity at Christmas or for a big promotion. After the busy period, the equipment is sent back, so the cost disappears.

Another method of turning fixed costs to variable costs is in the allocation of labour in the factory. The factory needs to be cleaned, which is part of being in the food business, and this cleaning roll is traditionally carried out by a hygiene team who are separate from the production crews. The factory needs cleaning whatever the activity level; you need to clean a machine whether it has made 100 products or 1000, so hygiene costs for labour and cleaning chemicals are considered fixed in a lot of food businesses.

A method of moving fixed costs to variable costs in a food business is to reallocate the cleaning task to production crews. If the factory is busy, the production crew makes product, if the factory is less busy, the manufacturing process is stopped and the production crew clean the machines. In this way the fixed costs of hygiene can become more variable as the production crews pick up the labour element of the cost, leaving only the chemical cost as a fixed one. It is obvious that the cleaning still needs to be carried out, and be paid for, but by moving the responsibility to the production management, it can be incorporated into the production plans more effectively.

In the old system, the hygiene team would arrive ready to start work but not be able to because the product crew had yet to complete the orders. The hygiene crew would keep themselves busy until the production was finished but would not be working effectively for that time. The new method would see the same crew arrive, and this time as production operatives, they would take over the production process (maybe stopping the original crew from incurring overtime payments) and continue production. At the planned time the production would stop and cleaning would take place. Production would then restart as soon as was required, with no need to wait for a new crew to arrive. In this way, food factory hygiene could be revolutionised and an eight-hour hygiene window in the schedule could be reduced. As a result of this type of thinking, the output of a factory could be substantially increased. Methods to improve the chemical costs of a cleaning process will be described later (see Figure 1.9).

Fixed cost reduction

Finally, the fixed cost has to be reduced or preferably eliminated altogether. Again the food industry has been working in this area for a long time and there are tried techniques to use. When you phone up a business you are often greeted by a computer saying: "press 1 for sales, press 2 for accounts, press 3 …" This automatic system has replaced the telephone answering duties of the receptionist, who used to receive calls and direct you to the correct extension. This is an example of a fixed cost, the receptionist being eliminated and replaced with a relatively low-cost alternative system. Obviously there may be an impact on the customer's perception of service, but this may be a decision that is deemed to be "worth it". In some companies, the receptionist still remains but now can be used on higher value work for the company. Fixed costs reductions are sometimes difficult to achieve; that, after all, is why they are called fixed costs. But remember, there is no such thing as a fixed cost.

Figure 1.9   Cleaning a production line – but is the cost fixed or is it managed flexibly to make the cost more closely linked to the activity level of the business? (picture courtesy of Ishida Europe)

Questions

How do you get the price down? The simple answer is to look in detail at each cost in turn (normally starting with the biggest) and ask a few questions:

- Do we need it?
- If we need it, do we need all of it?
- Is what we have the right specification?
- Can we find what we need cheaper elsewhere?

Questions

Before we go any further, take one of your main fixed costs, labour or materials for instance, and just consider your answer to the above questions. You may decide not to do anything differently (or it may not be your decision to make, but at least it will be an informed decision):

- Do we need it?
- If we need it, do we need all of it?
- Is what we have the right specification?
- Can we find what we need cheaper elsewhere?

Once these questions are answered, and they may take some time to investigate thoroughly, a decision can be made on the way to proceed. For example, imagine you currently have five forklift trucks at your factory.

Case study

Looking at a "fixed" resource and exploring options to reduce the cost

Your factory has five forklift trucks. Two are used by goods-in, two by despatch and one by the hygiene department to carry waste to the skips. Each of the trucks can be considered a fixed cost for the business, as they are all on long-term lease agreements, they are inspected and maintained on a regular basis, and they are all placed on charge for eight hours every day (assuming they are all electric trucks). If the business gets busy, they cope with the workload (but some run out of charge during a shift) and if the business is quieter, they are all still used, inspected, maintained and charged up. They are a fixed cost to the business. How can the fixed cost for our forklift trucks be reduced?

Questions

- Before you read on, have a think for yourself and see if you can come up with any ideas.

The simple way would be to call in the company from whom the trucks are leased and tell them that you want to reduce the cost and get them to come up with a proposal. That is a good example of Lean Thinking. "Why should I spend my time (and cost) thinking up a solution when I can get a solution cheaper elsewhere!" But, let's have a look at what might be possible. Using the four questions above, we might come up with a few options. The first thing we will need is information. How much is each of the trucks used during our busiest periods? How much is each truck used during our slackest periods? From this we could establish the minimum and maximum use of the trucks by the business. How about the maximum number of movements that are possible on one full charge of the battery? What is the minimum time to fully charge a battery? How many pallet and skip movements are projected over the coming days, weeks and months?

All information will take time to pull together and some of the numbers will be estimates rather than definite data, but it is possible to collect enough information to get a complete picture of the job that needs to be done. From this information, the answer to the first two questions will be:

- Do we need it? If we need it, do we need all of it?
- Do we need all of the trucks or could we manage with less? Could we manage for most of the time and only bring extra trucks onto site to help with busy periods?

We could also have information that would help with the third question:

- Is what we have the right specification?
- Are all of the trucks of the correct specification?
- Could we downsize trucks for use on lighter duties; would this reduce the price?

The final question is often the most difficult to answer; assuming we are comparing like with like, it is possible to work out a cost for the trucks that the business requires. This may mean trucks shared between despatch and goods-in or maybe that waste is only moved when trucks are available. But working out exactly what is needed, and then putting a plan together that matches the requirement, is a way of highlighting areas where fixed costs are too high and there is a potential for their reduction.

You will have noticed that during the whole of this case study we have ignored the fact that the trucks we currently have are the subject of a long-term lease arrangement. This was deliberate, as it is important in the process of the reduction of cost in a business that no boundaries are set early in the process. The aim of a lot of the techniques in this book is to release you from the constraints of the current situation and to develop ideas and solutions that are the "ideal future state". It is then a separate task to make that ideal state a reality in as short a time as possible (see Figure 1.10).

Figure 1.10   Forklift trucks are essential in a food factory, but are costly items to be under utilised.

Summary

The first chapter of this book looked at the structure and characteristics of the food industry and how it can respond to the pressures put upon it by its customers, consumers and the regulatory authorities.

As the theme is Lean Manufacturing for the food manufacturing industry, some time must be spent on considering why the industry is unique in the way that it needs to be managed.

There was a basic introduction to the theory behind cost and the areas of Variable, Semi-variable and Fixed Costs.

In the case study at the end of this chapter, some methods of lean thinking often applied to food manufacturing businesses was introduced. The aim of the fork-lift case was not to become expert in that aspect of running a food business, but to introduce some important concepts that will occur throughout the rest of this book.

Information is required for all decisions that are going to be anything but "gut feel". The collection of information about your business is one of the cornerstones in Lean Manufacturing. The final point in this chapter referred to the concept of the "ideal future state". It is an important part of Lean Manufacturing that no one feels constrained by what is the current reality, but they equally accept that to get to the ideal future state could be a long journey. They also know that in the food manufacturing industry, with all of its characteristics, pressures and food fashion, that during the journey the ideal future state will change, the goal posts will move and the journey will be subject to many diversions, detours and even the occasional roadblock.

Chapter 2

First steps to Lean Manufacturing

The overall aim of this book is to help people in the food industry to make improvements in performance. For things to improve, a recognition and understanding of improvement is required. This will ensure that the right improvement is chosen for the right reasons, obviously without sacrificing quality or safety!

Lean Manufacturing is a series of techniques that, if applied correctly, will improve the performance of a factory, a department or even a single production line or machine. The techniques of Lean Manufacturing are logical and as the techniques are applied, it can be thought of as a journey towards a more efficient future. Lean Manufacturing cannot be purchased and installed in a factory, the techniques being based on the way in which the work is carried out, and involves changing behaviours and attitudes of the personnel. You might like to read this chapter alongside the one on motivation.

Questions

- What behaviours and attitudes are you aware of in yourself and those around you that could affect your ability to introduce changes in your section/area? (e.g. think about people's reactions in the past when new things have been introduced).
- Have there been examples where you felt the change was not needed?
- Have people reacted against change because it was not their idea?
- What was the last change you introduced?
- How did people react to that?
- Why do you think they reacted in that way?

Lean techniques can be applied to all areas of a business and can give the business a great advantage against its competitors. Because Lean techniques cannot simply be purchased, they are difficult to copy. If you decide to buy a new more efficient machine or launch a new profitable product, your competitors will be able to do the same within a surprisingly short space of time. This is particularly the case in the food industry, where "copy cat" products can be launched within weeks of your new product hitting the shelves. Businesses are always looking to increase profitability and Lean Manufacturing can deliver extra profit in a way that your competitors cannot easily copy, because they are not

Handbook of Lean Manufacturing in the Food Industry, First Edition. Michael Dudbridge.
© 2011 Blackwell Publishing Ltd. Published 2011 by Blackwell Publishing Ltd.

working in your factory with your people in your way. Lean Manufacturing techniques are often based on "No cost or low cost improvement", so the use of the techniques can be applied by any company, no matter what the current situation with the finances. Indeed, Lean Manufacturing has been used to transform factories and to rescue businesses that have been in decline for many years.

If carried out correctly, Lean Manufacturing is an example of what is called a sustainable competitive advantage and will allow your factory to out-perform your competitors for a long period of time. But more than that, Lean Manufacturing can keep you ahead of the competition. The cornerstone of Lean Manufacturing is the development of continuous improvement systems. The factory is constantly striving to perform at a higher level. Even if it is at the top of the tree, the people in the factory are still looking to be better and better. This hunger to improve is a difficult thing to maintain in a factory and Lean Manufacturing has techniques that ensure that this happens. The better performance delivered by Lean Manufacturing techniques can be used by your business to do several things:

Extra output

Extra product can be made, because performance has improved and this releases extra production capacity. Chapter 1 showed that an increase in the volume of business can have a very positive effect on profitability, especially if the fixed costs can be maintained at their previous level. The extra product made will improve the profit margin. Of course, the extra capacity needs to be sold and not just used to increase stock levels in the business. The other way of utilising extra capacity in a business is use it as an opportunity to launch new products or get into new markets and broaden the sales base of the company. Finally, the extra capacity can be used to rationalise the production systems and to restructure. For example, older less efficient machinery could be taken out of use if other machines have increased capacity through the application of Lean Techniques. Shifts could be rationalised to reduce fixed costs, so weekend working could be stopped, allowing the costs associated with opening the factory up to be reduced or eliminated.

Lower costs

The improved performance will result in lower costs for the business and this could be retained and invested in your business or passed on to your customers in order to gain extra volume because your prices are now lower. Passing on the "cost advantage" to a retailer will hopefully allow them to gain more sales. The increased volume of production that results will mean that your business will be buying more materials and your ability to get reduced prices is enhanced. The extra volume will also allow you to dilute the fixed costs of the business. It is a circle of advantage that can be started with quite modest improvements in performance from the application of a few Lean Manufacturing techniques. Ultimately, in the food manufacturing industry, some of the factors by which the company is judged by others have to be guaranteed. Product safety is a given in the food manufacturing industry. Your company has to produce safe food or it will not be in business. Quality and service

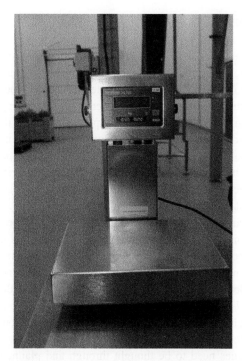

Figure 2.1   In order for improvements to be made to yield, data will need to be available. That data can only be achieved by the use of correctly calibrated weighing machines and detailed data collection. It is surprising the number of food companies who have an issue with yield from their process but do not have the ability to measure it routinely. It is a real chore to weigh every batch, but that is what must be done if yield issues are to be recognised, let alone solved.

level has to be very high level for the entire time. Occasional lapses may just be tolerable but consistent issues with quality and service levels will hurt the business deeply, as sales decline and the retailers look elsewhere for their supplies. That leaves us with cost as one of the few areas where companies can compete. The rest of the factors are just a given.

Better yields

The better controls in the factory will result in better yields and less waste and this too would result in higher profit margins for the business. There will be less waste because the machines do not break down as much as they used to. There will be less giveaway because the operatives now know what an impact that has on the business, they know it is monitored and they are motivated to get it right first time. Better yields because the goods reception area is now better organised and there is time to inspect all deliveries before they are accepted. Less waste from the storage areas in the factory because they are tidier and only contain what needs to be there. Finally, you get paid for your work because it arrives on time and in full at your customer, and you do not have to deal with whole consignments being rejected because they arrived late or there was a quality fault in the first box they opened at the depot (see Figure 2.1).

Improved systems

Better systems in the factory mean that changes can be made more quickly and this will make your factory quicker to react to demands from your customers; this, in turn, will increase the reputation of your factory and ultimately bring more business your way. For their important and high volume food items, retailers have a duel supply system where the product is sourced from at least two different companies. One way that factories can help a business gain extra volume is to be able to deliver at very short notice when the competitor factory has a problem. Quicker response cannot be at the expense of higher stocks or under-utilised capacity; those items would cost too much while you wait for an opportunity to use them. Quicker response should only come from having flexible systems that can be switched on, but that cost nothing when not required.

For example, your company is asked to increase output tomorrow because there has been a flood at one of your competitors' factories. Rather than have stocks of materials and packaging waiting for this opportunity, you should have responsive suppliers who are flexible and can meet your need to increase output. Rather than have spare machines waiting to be used, you should have flexible working arrangements with your staff that allow you to extend the length of the working day, compress the hygiene window and make the extra product that way. Rather than have a limited storage and transport capacity, you should have arrangements to move the product quickly so that space is no longer a limiting factor.

All of these techniques need to be thought through and planned in advance so that when the phone call comes in from your worried customer, your business can say "no problem – leave it with us."

The final aspect of better systems is that your business needs to be flexible both ways, not just able to increase output at short notice but also to be able to reduce output quickly without losing control of costs. It only takes a patch of bad weather for the major retailers to reduce their orders. Unless you have systems that allow you to reduce your costs in parallel, your factory could be running at a loss for several days. A few lost days can take weeks or even months to recover in a business that works at such low profit margins.

Better working environment

Finally, the use of Lean Manufacturing techniques will make your factory a better place to work. Workers will be more in control of the process of meeting very high performance and delivery demands and are able to do so consistently. Everything will be more organised and less frustrating, it will be easier to get things done and it will be a tidier and safer place in which to work (see Figure 2.2).

Result!

If you think that Lean Manufacturing techniques may help your business to be more successful, then how do you get started?

If you want to improve something – first you have to measure it.

Figure 2.2 Using Lean Manufacturing techniques can improve the workplace organisation – everything in its place (picture courtesy of Ishida Europe).

The statement above means that making improvements in a factory is a process of changing things to make them better and then knowing if the change has been a success? The only way of knowing for certain that an improvement has been made is to measure before and after and compare the two numbers.

If you want to alter a machine to reduce the quantity of waste, measure the amount of waste, make the adjustment to the machine and then measure the amount of waste again to see if there is a significant improvement.

Yes!	**No!**
Well done, you have made an improvement.	Oh well, it was worth a try. What else could we do to reduce the waste?

There is no way you can be sure that an improvement has been made without taking an initial measurement. This first measurement will also provide a record of the journey, so that in a week or a month, the improvement in waste can be checked again to make sure that the improvement has not been lost.

From the example of waste described here, it is simple to see that measurement needs to be applied to all things that which need to be improved.

Some techniques will be described later, which allow us to choose what to measure and what not to measure. However, for now you can assume that if you want to improve something, first you have to measure it (see Figure 2.3)!

Figure 2.3   A typical food production line, containing machines and systems and many opportunities to improve performance. This line is at the National Centre for Food Manufacturing and is used to train people in Lean Manufacturing techniques.

Assuming you already take many measurements in your factory, some may be for legal reasons such as the average weight of the packs that you have made. Some measurements will be to make sure the payroll is correct or to control yield such as counting the quantity of product you have made so that your customer is satisfied.

Measurement is required throughout a factory; the trick of Lean Manufacturing is to make sure that the effort, and cost that goes into measuring, is well spent. Measurement needs to be well targeted to provide information on performance, and for that information to be widely known and used in the business.

A factory that has implemented Lean Manufacturing techniques looks at the cost and value of everything in the business. If the cost of the information is larger than the value that the company gets out of it, then a change is needed to improve the situation. Later this book will show how Value Stream Mapping (VSM) is applied to food processes and why it is also applicable to the process of measuring within the factory.

Summary – First steps to Lean Manufacturing

This chapter has shown that Lean Manufacturing is about the continuous improvement of a factory's performance. It is the management of this process of change that is key to the use of these techniques.

If you want to manage something, first you have to measure it. This is the key message to carry forward to the rest of the book. Without measurement, nothing can be managed and no improvement can be developed and implemented successfully.

   The final thing to remember is that it is people who carry out Lean Manufacturing techniques. The management of the people in the factory is a high priority in Lean factories. Lean Manufacturing is about making improvements by a continuous chain of tiny steps. To make Lean Manufacturing a sustainable, competitive advantage will take the effort and contribution of as many people as possible to improving their bit of the complex jigsaw that is your factory. The low cost and no cost improvements required to make factories better places in which to work need to be implemented in a way that keeps the staff in the business hungry for the challenge of the next improvement.

Chapter 3

Teamwork and the development of solutions

Understanding the principles of Lean management is one thing and you can probably begin to see how they would benefit your factory. The next thing to consider is how you can engage your team in the process of making Lean Manufacturing a reality on your production line, in your department and in your factory.

Questions

Think about your team:

- What challenges do you imagine you might face, or have faced, in trying to implement something new?
- How do you currently get the team involved with new ideas?
- What difficult attitudes or responses have you come up against, if any?

Sometimes being a team leader, a supervisor or a manager can feel as if all the responsibility and decisions fall on your shoulders alone and if it were not for you, nothing would ever get done... Sounds familiar? As with the principles outlined in the previous two chapters, of looking at using the resources you have in the most efficient way, effective people management encourages you to do the same. How can you use the people resources you have, including their ideas, their contributions and their reservations, to create the most productive factory you can? This chapter will help you begin to think about this and will give some ideas about how you might use your team more effectively. Lean Manufacturing is about a process of continual improvement and big changes can result from just one big leap forward, but that leap carries with it a level of risk. In Lean Manufacturing the same impact can be made on your business by taking a thousand tiny steps, hopefully all in roughly the same direction. This way of making change is lower risk, with usually no cost or low cost, and can be equally as speedy. But coming up with thousands of ideas is not an easy task for one person. Ideas and the ability to carry them out are best carried out by teams of people working together.

Handbook of Lean Manufacturing in the Food Industry, First Edition. Michael Dudbridge.
© 2011 Blackwell Publishing Ltd. Published 2011 by Blackwell Publishing Ltd.

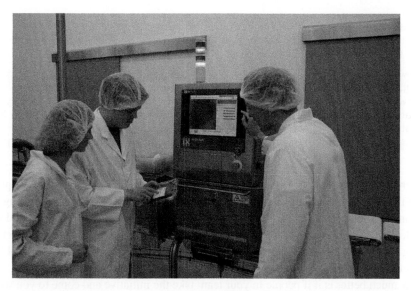

Figure 3.1 A quick meeting on the shop-floor is maybe all that is required to find a solution and make an improvement.

In Lean Manufacturing, one plus one equals more than two, sometimes just a little bit more that two, but it is difficult to always come up with ideas to save 50% of the waste. Coming up with an idea to save 0.1% of the waste is an improvement and when added to the other 500 ideas for waste reduction, will add up to the same as the one big idea!

Solutions, not problems

If you are the sort of manager who feels as if they are in *Groundhog Day* answering the same questions, day in and day out, then this is for you. This will look at how you start the process of getting people in your team to begin to come to you with possible solutions rather than just problems that they need you to fix.

For example, every morning at the start of the shift, one of the machines does not work because the electric panel is full of water. The team leader comes to you moaning about the hygiene team and you say you will have a word with them. First though, you call the electrician, who is also getting fed up with the calls every shift, to come and dry out the panel and get the machine going. Now, this situation could continue for weeks, causing inefficient production runs because of the lost time. You could spend time sorting out the problem but you are busy with other stuff.

A solution to this is to get the people with the correct skills together to come up with a solution. Maybe if the hygiene operative, the production team leader and the electrician met for just a few minutes, they could come up with a solution (see Figure 3.1).

How can you begin to create the climate where that is more likely to happen than not?

Questions

First, think about your style and approach:

- How would you describe it?
- Is that what you would like it to be, rather than what actually happens in reality?
- How do you think others would describe it (your manager, your colleagues, your team?). Do you know? It may be worth checking out (see Appendix 1 for an easy format for gathering feedback about your leadership style).

There is a saying that "behaviour breeds behaviour", which means that people partly create the sort of response that they get back. If your team wants to come up with suggestions and possible improvements, yet never have the chance to try them out or be told why they do not work, then they will stop bringing their ideas to you. And then you get left feeling frustrated when they continue to ask you things you think they should be able to sort out on their own.

How much better is it if people in your team take the initiative and come to you with a solution? Especially if they are coming up with solutions to their own issues. They will feel ownership for the idea and they will be motivated to make it work and make sure it keeps working. Any idea that you come up with may have the "not invented here" tag attached and the idea may not be sustained because you cannot be there the entire time to sustain it.

Obviously there may be very good reasons why everything cannot be tried out. When working in a food factory, rules have to be followed. The idea about reworking waste product back into the mixing bowl might not be possible for very good technical reasons, but a response to the idea should encourage workers to come up with more ideas and help them understand why that particular one would not work. There does need to be a system of control over the ideas and improvement activity in the factory. Ideas need to be checked out before they are implemented or the team may end up causing more problems than they are solving. The checking out of ideas can be done very effectively if the improvement team who come up with the plan is made up of the right mix of people.

Looking again at our "wet electrical panel" team from earlier, the team is made up of the team leader for the machine, a hygiene operative who cleans the machine and an electrician. In that group there are all the skills and knowledge to come up with a no cost or low cost solution.

This is what may take place at their meeting:

Electrician – We have got to stop using so much water to clean this machine; it gets everywhere.

Hygiene operative – I am given 20 minutes to clean this machine. I have to use foam and rinse it off to get the job done in time.

Team Leader – All I want is for the machine to work when we switch it on. Can we improve the seals on the panel to stop the water getting in?

Electrician – The seals are okay if you do not hit them with the full force of a pressure washer at more than a 5 cm range!

Hygiene operative – It is the only way I can shift the muck from the front of the panel.

Team Leader – Okay, here is a solution. I will get the production guys to wipe the panel during the shift so that it stays clean (they should be using "clean as they go" anyway) and sanitise it at the end of the shift. The hygiene team can then avoid cleaning the panel and protect it with a plastic cover when they foam and rinse the rest of the machine. I will check with the boss to get a plastic cover made and tell him that I have just spent £3 of his budget (see Figure 3.2).

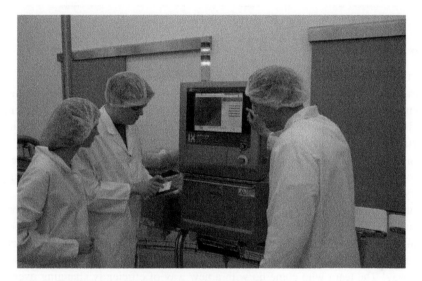

Figure 3.2 Solution found in 3 minutes, with only £3 spent.

Encourage them to come up with a suggestion or idea.

Also, the answers may be obvious to you because of your experience and the fact that you have seen, and dealt with, similar problems before, but how did you learn and gain that expertise? You learnt by example, or by trial and error, or by watching and asking someone who knew. Part of your mission is to free up some of the time you spend responding to questions and having to continually check that the right things have been done correctly.

It is one thing to say "Of course you can come to me if you are not sure…" or "my door (if you have one!) is always open," but if you are rushing around looking stressed, then the message that is received is something like "don't even think about it!" Rather than come and discuss something with you, people may tend to just get their head down and muddle through or give up and take the attitude of "oh well, they're the boss…"

How can you signal that you welcome ideas and suggestions from the team? When is a good time for them to catch you? How can you encourage all the people working in your area to think about the challenges that are being faced?

Questions

Write down some thoughts of how you could do this:

- Signals that you like new ideas
- Time when it is best to talk with you
- Encouragement of people to think about improvements

Make time

Make time, either informally by going round and talking to people, asking them what is working and how they think it could be improved or more formally, say at the end of your morning briefing. Maybe make a tour of your area at the start and/or end of your day. Talk with people about the issues and let them know that their opinion is important to you.

Think about too how you handle mistakes or how you respond when someone else does something wrong. Obviously it will need to be corrected but if the response you give is angry, over-critical or ridiculing, then people will fear putting forward alternative ideas for fear of criticism or being seen as stupid. But remember every new innovation had to be thought of by someone who saw a different way of doing something. The first person to think it was a good idea, in the 1980s, to manufacture pre-made sandwiches, created a whole new section of the food industry. They were invented in Northampton, in the UK by the way! The team who come up with a new way of efficiently changing over your production line from one product to the next could revolutionise your business. The ideas are already within your business part of your job, and one of the features of Lean Manufacturing techniques is to get those ideas out from where they are hiding and get them implemented in a controlled way.

How do you encourage people to think of solutions, rather than just state what the problem is and expect you to sort it?

Think about these two conversations

Bill: Look, this machine isn't running properly – it's all lumpy! What shall I do?

You: Adjust the swing arm, get the machine speed correct and check the temperature. If that doesn't work, do call me back and I'll sort it out for you.
[while thinking …but they should know that! Weren't they paying attention when we covered that the last time?!!]

---

and

> **Bill**: Look, this machine isn't running properly – it's all lumpy! What shall I do?
>
> **You**: Okay. What options do you think you have? What do you remember from the training session we did last month?

There are obviously times when the first response is the best one, such as in situations that are safety critical for instance, but whilst it sorts the problem in the short term and seems like it is quicker, it does not:

- help Bill understand what to do next time
- check whether he actually knows or just is asking you because it is easier
- give you a sense of whether the training worked/whether Bill attended
- indicate to Bill that you expect him to sort it out or that you see it as part of his responsibility.

The more you can ask questions to get them to think something through, the easier it is to get them to know that you expect them to be thinking, wondering, remembering; in fact playing an active part in helping you to improve the performance of the factory!

> **Questions**
>
> Things to consider when asking problem-solving questions are:
>
> - Is this task new to them?
> - Have they experienced something similar?
> - Have they been around others doing the task/dealing with this situation?
> - Could they have a "best guess"?
> - Is there only one way of doing this?

It has now been shown that your team will behave in the same way as others that they see around them and that a lot of their behaviour is a direct copies of their boss's behaviour.

How do you get someone to care enough about something to come up with an improvement? The first part is to show that you care about it; show that it is important to you. The second part is to make sure that each person is part of a team that cares and shows it cares.

These things are easily said but not so easy to achieve, especially if the team has been established for a long time.

Lean Manufacturing is not easy for a competitor to copy, so why cannot you just buy Lean Manufacturing off of the shelf and install it in your factory, in spite of what some consultants may tell you? Lean Manufacturing is as much about the way that people work together, their values and beliefs, their attitudes and behaviour, as it is about the application of a technique or tool. Perhaps it is clearer now why the implementation of Lean Manufacturing is a journey to be travelled. People take time to change. Teams take even longer to modify their behaviour.

Teamwork has many advantages

- A greater variety of complex issues can be addressed by pooling expertise and resources. Teams are good at quickly exploring all avenues around an issue. The blind alleys are quickly identified and the ideas that show promise can be refined into a workable solution.
- Problems are exposed to a variety of knowledge, skill and experience. It is important that the teams are constructed to make sure there is a balance of knowledge, skill and experience but also of personality types if possible. Some people are great at coming up with an innovative solution to a problem but that same person might not be good at the fine detail that is often required to make the idea work in practice.
- The teamwork approach boosts morale and ownership through participative decision-making. Because the team came up with the solution, they have an "investment" in making it work. The celebration of success is too often forgotten in a busy factory. The recognition of a piece of improvement activity will encourage the members of the team to repeat the exercise. The recognition is often not a financial one and non financial incentives will be looked at later in this chapter.
- Improvement opportunities that cross departmental or functional boundaries can be more easily addressed. Often a change in a food factory will have implications outside of the immediate area. It is important in the development of the improvement idea that all of these implications are considered. It may be decided that some of the implications are a price worth paying to get the improvement done or it may be that the knock-on effect of a change is too great and the idea has to be scrapped.
- The recommendations are more likely to be implemented than if they come from an individual. The power of a team in securing the support of senior managers cannot be over emphasised. It is vital that support for a change is received in order to send the signal that "yes, this is the sort of thing that this factory needs to be doing." Again this will start to change people's behaviour and their attitude towards change (see Figure 3.3).

The issue of teamwork has been written about and discussed by managers for many years. Most of that work has looked at the establishment of a team to work together on a particular task. Most consideration has been given to a team that is long-lived and has had a chance to develop.

Lean Manufacturing in a food factory is trying to install a system of continuous improvement that will revolve around the establishment of an improvement team. The main feature of improvement teams is that they are put together to look at a particular issue and to work rapidly towards a no cost or low cost solution. By their nature, some improvement teams are very short-lived and the projects are carried out very quickly. It is this speed of action that allows Lean Manufacturing techniques to transform the performance of a business very rapidly.

How do you get your team members to work well in a team that will only be in existence for a few days, or even a few hours? The answer comes in how these teams are managed and how their work is recognised.

Figure 3.3 Teams of people working on a task are capable of achieving higher performance than individuals. They are able to discuss techniques and systems and come up with better ways of doing things. A team needs to be correctly managed to get the best out of them and ensure they are all contributing as much as they can to the team performance.

Management of short-term team-working

In order for team members to work well together, from minute one of its existence, they need to be very clear about the team objective and their individual roles within the team. Without this clarity it will take too long for these roles and objectives to emerge and the response time will be too slow. The team will need a delivery target to drive them into action. The management of these short-term improvement or action teams, in this way is built into Lean Manufacturing management systems that will be covered later in this book, but for now it is enough to say that the control and driving of improvement activity is a major part of a production manager's life in the food industry. It is not just about achieving today's production plan within budget; it is also about managing the future performance of the factory, its equipment and the people in it.

Recognition of achievement

Lean Manufacturing is everyone's role within the business and one of its cornerstones is the measurement and the continuous improvement of performance. Often at the start of a Lean Manufacturing activity in a business, the buzz about the building is enough to get the ball rolling. But how do you keep the ball rolling and how do you make it accelerate?

The initial enthusiasm for Lean Manufacturing in a business is often caused by it being looked at from a senior level. People from the senior management team are seen on the shop-floor and suddenly the workers feel responsible for the delivery of something other than the boxes of food they make each day. They feel as if it is their time in the spotlight. Things start happening and there is an air of excitement.

TWIDDLE cc   Pay Slip 2200 : Period ending 02/08/2006

Hours: 27/07/2006 – 2/08/2006

Employee: Z MDUNGE (ZACHARIAS)

Pay number: G07	ZACH	
Occupation: GENERAL		
Income Tax Number:	ID:5411110065086	

PO BOX 1742
SELBY
JOBURG
    2001

Employer UIF number: U11111111

Employer UIF contribution	5.42
Accumalated overtime: Before	2.93
Accumalated overtime: Now	2.93

TIME WORKED:	Hours	Tariff	Amount
Time Due	45.00		
Fixed Wage	41.00		400.00
Weekday Overtime	2.10	13.34	28.00
Short time/ unpaid leave	4.00	8.89	−35.56
Total			**392.44**

Leave taken during period:	Start Date	End Date	Days within pay slip
Sick leave	01/08/2006	8/1/2006	

WAGE CALCULATION:

Total wages due (as set out above)	392.44
BONUS	150.00
FUNERAL	−5.00
Food Deduction 2	−8.00
UIF	−5.42
Total payable (Method of payment: Electronic)	**524.02**

Signature _____

Date _____

Figure 3.4 A payslip containing a productivity bonus scheme – this is one way of recognising a contribution to the company performance, but it is not the only way.

After a short period of time, it is often the case that the momentum starts to drop and the impact of the improvements starts to slow down. This is as a result of the "easy targets" or the "low hanging fruit" worked on in the first few weeks and by now the challenges are getting a bit more difficult to meet. It is usual at this time, as the level of effort has to go up, that people start to ask "what's in it for me" or "why should I come up with further systems to improve productivity or reduce waste, when I get nothing for it" (see Figure 3.4)?

It is at such times that some companies tackle this issue with a rewards scheme that allows people to benefit from the improvements made. There is a danger in the introduction of a bonus scheme that you could stop the activity that you are trying to encourage. A bonus based on coming up with ideas, sometimes called a suggestions scheme, rewards the idea but not the large amount of hard work that needs to go into its refinement and implementation. A financial bonus scheme based on production line performance will not recognise the efforts made by people who are working off line or on a new product that is struggling to meet its technical specification and is generating high levels of waste.

The construction of a rewards system for your factory is a specialist area not covered here, but the recognition of achievement is vital to the continued momentum of Lean Manufacturing journeys. This is often addressed in factories by a system of non-financial recognition; ways that the company recognises the achievements of an individual or team. The type of recognition varies from business to business. In its simplest form, it is a "thank-you" from the boss, verbally, but even better in writing. Other recognition schemes have "employee or team of the week" at their core. At the end of the day, what this does is give the people in the business the recognition that they often seek and as a result they will engage fully with the improvement activity when the opportunity arises.

Summary – Teamwork and the development of solutions

This chapter has looked at the role that people can play in a system of Lean Manufacturing. It is the behaviour and values, personality and motivation of people, which is the difference between successful and less successful factories. Of course, the equipment plays its part too, but having the best equipment does not guarantee success; having the best people maybe does. The best people will improve the performance of the equipment. They will design things so there is less waste, and will develop systems to ensure that all key areas of the business are performing at the highest level.

This chapter has also looked at some of the issues of working in teams that are short-lived. How can a team perform well if they are together for just a few hours?

Finally, the issues around recognition were addressed and the sticky problem of financial and non-financial rewards.

Chapter 4

Starting to measure and quantify performance

"What sort of shift are you having?" says the boss.
"Not bad," you reply, "Not a bad day so far."
Compare the response you gave to what might have been.
"What sort of shift are you having?" says the boss.
"We are 12 minutes ahead of schedule, our waste is 0.8%, which is 0.2% better than target and we have a labour efficiency of 102%. We had a delayed start because of water in the electrics, but the engineers got us going with only 6 minutes delay and the engineers are coming back on the next changeover in 22 minutes to seal the leak in the electrical panel."

Which do you think the factory manager would prefer to hear?

Questions

- What sort of response would you be capable of giving to the question?
- What would you need in order to improve the information you have at your fingertips during the shift?

The second response can only be made if a system of data collection and controls are set up to keep everyone informed of the performance of the process of a department or production line. If a system can be set up and managed, it allows the production process to be managed in an efficient way to ensure that performance is optimised (see Figure 4.1).

Questions

So what measurements do you take in your business?

- How are they recorded and communicated to the people who need to know? Think about:
  - Output
  - Labour usage
  - Material usage
  - Waste
- What else?

You might notice that the measurements you are asked to take in the factory are very similar to our lists of variable costs that we looked at earlier.

Handbook of Lean Manufacturing in the Food Industry, First Edition. Michael Dudbridge.
© 2011 Blackwell Publishing Ltd. Published 2011 by Blackwell Publishing Ltd.

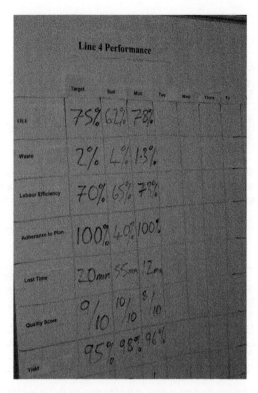

Figure 4.1   Here is an example of the kind of measurements typically made in the food business to monitor performance. You will notice that each measure has a target value for everyone to aim at.

The system of KPIs and SIC

There are two systems to control fast moving production processes that are typical in the food manufacturing industry. The first is called Key Performance Indicators (KPIs).

A KPI is a measurement that is taken and recorded, to ensure that the important areas of a process, from a performance point of view, are on view to people who need to know. A KPI could be anything about the process and the people operating it, which could have an impact on performance.

KPIs can be developed around almost any measurement and then be used to indicate the overall performance of the line. It is normal that a KPI is a combination of two or more factors rather than a simple measure. In this way it relates two or more factors into one number so that the value of the KPI is more comparable day-to-day, week-to-week, even though the business may be changing in that time.

For example, a KPI that is used to measure the output of a process could be simply the number of packs made in that shift. If the output of the factory varies for different manning levels, then a more useful KPI might be the output per man hour. Output per man hour would allow you to compare like with like for days where the factory had 150 people

Waste KPI for the week Production Line A			
Day	Output	Waste	Percent Waste
Monday		652	
Tuesday		642	
Wednesday		687	
Thursday		623	
Friday		612	
Saturday		452	
Sunday		398	

Figure 4.2 The quantity of waste created each day shows that Saturday and Sunday are the days producing the least waste during the week. By counting the black bags and weighing the skips, there would be an investigation into why the waste on Wednesday was the highest.

Waste KPI for the week Production Line A				
Day	Target	Output	Waste	Percent Waste
Monday	4.50%	15004	652	4.30%
Tuesday	4.50%	16052	642	4%
Wednesday	4.50%	16542	687	4.20%
Thursday	4.50%	14882	623	4.20%
Friday	4.50%	16017	612	3.80%
Saturday	4.50%	8452	452	5.30%
Sunday	4.50%	7452	398	5.30%

Figure 4.3 By adding the production output quantity and by doing a simple calculation to express the waste as a percentage of the output, it can be seen that Wednesday, far from being the worst day of the week, was actually a normal day. It was Saturday and Sunday where the waste got out of control compared with the output of the production line.

Waste KPI for the week Production Line A				
Day	Target	Output	Waste	Percent Waste
Monday	4.50%	15004	652	4.30%
Tuesday	4.50%	16052	642	4%
Wednesday	4.50%	16542	687	4.20%
Thursday	4.50%	14882	623	4.20%
Friday	4.50%	16017	612	3.80%
Saturday	4.50%	8452	452	5.30%
Sunday	4.50%	7452	398	5.30%

Figure 4.4 One of the most important aspects, the target, has been added in the use of KPIs. This is the number that the business has decided on, at least for now, to be a realistic aim for the people working on Line A. The only two days where the target was exceeded were on Saturday and Sunday. Of course, as improvement work is carried out, this target could and should be lowered.

and was very busy, to a time when the factory was having a quiet period and only 90 staff were employed. In the same way, waste could be (and often is) a KPI. Just recording the weight of waste produced could be misleading. Look at the table above to see how KPI can be used to give a true view of performance (see Figures 4.2 to 4.4).

Because almost any measurement can be taken, it is possible to disappear under a mountain of information and data and as a result fail to spot an important indication that there is something wrong with the process. For this reason, the *key* piece of the KPI is that the information is *key* to the business but does not attempt to measure everything. KPIs are, as their name suggests, key pieces of data and information that give a strong indication to the overall performance of the process being measured.

What do we need to measure – what needs to be our KPI?

Questions

- What are your KPIs?
- Do you have any?
- Do you know what you have to do to be able to improve them?

The first rule of selecting KPIs is to remember that they can be changed at any time if it is felt that they are not helping with the process control and improvement. The KPIs set up on a new production line may be changed at some point to focus the production team on new targets and matters of concern, or a new KPI can be developed if pressure on the business changes and suddenly priorities change.

For example, the cost of a raw material goes up steeply and its yield needs to be more closely controlled. A KPI could be developed to closely monitor the yield and waste on that material and ensure that all staff are aware of the performance in this area.

Another example would be that a machine performance was causing concern. The number of breakdowns appears to be increasing. A KPI could be developed to monitor and control this situation and ensure that a solution was found. The KPI could be the number of stoppages of over one minute that occurred during the shift, but maybe a better one would be the Mean Time Between Failures (MTBF). MTBF is the average length of time that the machine runs before the next breakdown. As improvements start to be implemented by the improvement team working on the issue, it will be very apparent that the MTBF is increasing.

It is important that the number and type of KPIs used are monitored. Too many KPIs lead to excessive paperwork, and the wrong KPIs mean that issues and opportunities to improve can be missed. Too few KPIs can lead people into the "not a bad shift" view of performance rather than the focused system that is required if Lean Manufacturing techniques are to be applied. It is the local team who should agree the KPIs by which they wish to be measured. The KPIs are their tools to keep a close watch on their performance and how that performance is changing over the weeks and months. The KPIs can, if correctly selected, also be used to compare performance between different lines, different departments and even different factories. This is a process called Benchmarking that is important enough for another look later in this book.

The other vital part of any KPI is to have an agreed target for the measurement, so that everyone knows what it is expected to be. The target can be set in many ways but it should always be realistic, achievable but stretching, and be linked in some way to the whole business.

For example, a target of zero waste for a food production line is a desirable one. but there are very few processes that are capable of achieving this target. Better to set the target based on recent performance and maybe the level of waste that is built into the product costings.

Waste will be further discussed in other sections of this book, as waste control and yield are vital factors in the food industry, to ensure that costs are controlled and the very small profit margins of food manufacturers are protected.

Setting targets

Every KPI must come with a target. The target should act as a motivation for the people who are working to achieve it, rather than be an out of reach/depressing target that no one believes is possible. The targets set should be communicated clearly to everyone, so that they are clear about what is required and they have a view of "what good looks like". It is easy to lose sight of what is to be achieved when focus in the day-today operation of a food factory moves away to something that is a hot topic. For example, it has been observed many times that the performance of a production line will drop when a new production line is installed in the same area. The drop in performance is not deliberate but people spend time focusing on the new line, and the established line, the old workhorse, is seen as in some way less interesting. The focus of the people is distracted and this needs to be managed before performance drops too far. Careful monitoring of the KPIs and comparisons with the targets will soon highlight any drop off in performance so that corrective actions can be taken.

It is usual for KPI targets to be revised every year, but more often is possible especially if there has been a lot of improvement activity and the targets are no longer stretching and are met without fail. The KPI itself can be introduced and dropped in response to the requirements of the business, but it is usual to have core KPIs that never change and then some other KPIs that change on a regular basis. Typically, core KPIs focus in the major cost elements of the process and so would contain a KPI for:

Production efficiency	How much was produced compared to how much could have been produced in the same time, had everything run perfectly.
Labour productivity	The output per man hour.
Material yield	How much was used compared to how much should have been used to make the product.
Product quality	How much was made right first time compared to total output.
Availability	The percentage of time the machine could run compared to the amount of time the machine was required to run.

See Appendix 2 for the calculations behind each of these KPIs.

All KPIs must have a target. Targets can be changed at any point but it is clear that if a target changes too often people will become disillusioned and will stop trying to meet or beat it.

"We worked really hard to hit the output target KPI for today – now the efficiency target has changed and so tomorrow we have to do more output in the same time."

When targets are changed it should not be imposed on the team; they should agree that the new KPI target is achievable and they should sign on to finding ways of meeting it. The team will need time to find and develop techniques to hit the new target and this should be a focus of activity and managerial support for a period after the KPI target has been changed. The secret of the use of KPIs to monitor the performance of your factory is that someone must own the KPI. They must understand it and know the factors that make the KPI move. Someone must have their hand on the levers of each KPI in a factory, if they are going to be able to help in improvement activity.

Here are some typical KPIs that could be employed to aid you in the monitoring and control of your process, production line, department or shift. It needs to be remembered that the role of a KPI is to monitor performance. A KPI will not take the corrective action necessary to make any changes; that is down to the management of the factory.

People KPIs

% of absenteeism

That is, of the people who were expected to come to work on this shift, how many failed to show up. Absenteeism is a major disruption in factories and monitoring it can help in its management.

% of operatives trained in machine operation

This is a useful KPI, as it is easy to measure and gives an indication of the ability to be flexible and react efficiently to situations. Having strength in depth is like a football manager having a super substitute on the bench waiting to come on. If the usual machine operator is not available, will there be someone who can carry out the role to the same level of performance. The reason your machine operator is not available can be anything from holidays to sickness, but it could also be that the machine operator is standing in for the team leader, is on a training course or is taking part in a short meeting about an improvement opportunity or a new product launch. Strength in depth will mean that all of these activities can take place without a loss of performance on the machine.

% of operatives trained in a Standard Operating Procedure (SOP)

This KPI gives a view of the level of knowledge in the production team. SOPs are discussed later in this book.

% of operatives trained in a HACCP monitoring task

Hazard Analysis Critical Control Points (HACCP) is the way in which Food companies control the safety of their products. There are points in the process that have to be carefully monitored and controlled to ensure that the food produced will be safe to eat. The percentage of operatives that are trained in the monitoring and control of the HACCP areas of the process is an indication of the ability of the company to respond to events without seeing an impact on the business.

% of operatives failing hand swab tests for insufficient hand washing

This is an interesting KPI and indicates directly the issue of hand cleanliness in the factory. But it can also be used as an indication of general levels of discipline among the operatives. A person who fails to wash their hands correctly on their way into the factory believes that the rules do not apply to them! That person is likely to have the same attitude towards other rules in the factory. The KPI does exactly what it says on the tin. It is an indication of some of the performance factors in your business. When something is difficult or impossible, or too costly to measure directly, a KPI can be used to keep an eye on that aspect of the operation. Measuring people's adherence to the rules might be difficult and would be open to interpretation of the observer. A hand swab failure is an indisputable fact and so has the advantage of being an indicator of adherence to rules by the operatives.

% of ....

There are many potential KPIs that focus on the behaviour and attitude of the people in a factory and they allow managers to monitor the situation. As with all KPIs, knowing about it and having the ability to do something about it are two separate things. Getting your hands on the levers of a situation and having the ability to make something change is the main thing that distinguishes an effective manager from one that is less effective in the role. One technique to ensure that the levers of control in a factory are used to make an improvement in the KPIs is to ensure that responsibility for a particular KPI is allocated to an individual. It becomes their responsibility to monitor, control and improve performance in that area. In this way the KPI becomes the measuring device that is used to record progress and improvement.

Once you have got some measurements, the next step is to identify what needs to be controlled to keep the KPI within the acceptable levels, or even improve it.

These are the levers of control in your factory; what do your levers look like? How would you improve the absentee percentage for your team? What lever would you pull to ensure that the KPI of hand swab failures improved?

Questions

- Do you have any other KPIs that are focused on the people in your business?
- How about percentage of people making an improvement suggestion in the last month?
- You might have some based on lateness or disciplinary records.

Machine KPIs

% Downtime due to breakdowns

This is a KPI that is often used by production teams to monitor the performance of the maintenance function in a factory. It looks easy at first; measure the time that the machine is stopped when you wanted it to run, but who decides what a breakdown is? Who feels responsible for this KPI? An engineer arrives at a "breakdown" to find that a guard has not been closed properly on the machine. An engineer arrives at a breakdown to find water in the electrics. An engineer arrives at a machine to find an emergency stop has been triggered accidently. All these are potentially breakdowns but who will monitor the KPI and take action if the percentage of downtime due to breakdown is not improving. Is it the engineer, the machine operator, the hygiene operative or someone else? All that is clear is that someone needs to be made accountable and responsible for the percentage of downtime due to breakdown.

Average running speed

This can be a useful KPI and easy to calculate, so can be used to look at performance over a very short period of time. The average running speed is simply the quantity of product made or packed per minute or per hour.

For example, a packing machine is targeted to run at 100 packs per minute, which are 6000 packs per hour. The shift starts at 06:00 and by 07:00 only 4200 packs have been made. This indicates that either the machine is running slowly, or the machine has not been running for the whole hour because of delays or breakdowns, or the machine has been running at the correct speed but has made a large number of rejects. Later in this book it will be shown that this type of analysis will help highlight machine and factory performance in a system called Overall Machine Effectiveness (OEE). In OEE, the under-performance of a machine is split into:

- **Availability** – the machine was supposed to be running but was not.
- **Performance** – the machine should have been running at 100 packs per minute but was running at a different speed.
- **Quality** – the machine was running but making rejects.

Average Running speed is a simplified version of OEE, but if used as a KPI, could indicate where improvements could be made.

Average changeover time

In the food industry it is rare for a food production line to just make one product day in, day out. It will be necessary for the production line to be changed over to make a different flavour or size of product. Some production lines may change over once a shift, some may change over once an hour. The changeovers need to be conducted as efficiently as possible, to minimise the downtime and therefore reduce the loss of production time. A KPI of average changeover time is a useful one to monitor the performance in this area. As with all KPIs, it needs to be clear how the KPI is to be measured. In this case, changeover time is usually measured at one point on the machine, say the exit of a tray sealer, and the time starts when the last product exits the tray sealer and stops when the first good product exits the tray sealer. Once the changeover time is captured, work can begin to identify ways of reducing the time. The improvement work will be monitored by the KPI so that the progress can be recorded and the improvement team, and everyone else, can see the results of their work.

% of downtime due to lack of materials

This is a good indicator of the level of communication within the factory. If a production line is unable to work because of the shortage of materials, it indicates poor communication between the line and its supplier. The supplier may be another line; it may be a warehouse or stores function, or maybe another company who are late with a delivery to your site. Occasionally a level of disruption in the supply of materials can be expected. This can be for a multitude of reasons. Lorry breakdowns, machine breakdowns and even that you need more than you first thought. There will be a reason for all material supply failures; some can be managed, some cannot. By using this KPI, the incidence and severity of this disruption can be monitored and the effect of any corrective actions can be seen. One important improvement that could be made is in the communication of the failure before it happens. If your production line is given some advance notice that there is a problem with material supply, then decisions can be taken to minimise the impact of the disruption. This could be everything from shutting the line early for a lunch break to changing over to a different product. With communication comes the ability to minimise the impact of a problem.

% of downtime due to late back from break

The KPI is an example of one that may be created to monitor a particular issue that has been highlighted. The very fact that people know that this situation is monitored may be enough to correct their behaviour.

Number of unplanned changeovers per shift

Again, this is one of the KPIs that can indicate several things to the managers and the teams. The number of changeovers should be minimised to keep machinery manufacturing product as opposed to being idle. The number of changeovers can be an indication of the complexity of the operation. A business in short shelf-life food with a large range of products will incur more changeovers; that is inevitable. What this KPI can indicate is the number of unplanned changeovers that have occurred. Changing over a production line is sometimes used as a way of reducing the impact of a problem and therefore can be used to hide problems in the factory. The KPI will keep an eye on this tactic and ensure that the root cause of the problem is not hidden simply because its impact was minimised by some quick thinking on the shop-floor.

For example, an unplanned changeover may come from a material supply problem. But an unplanned changeover may also be the result of a quality problem, where a product has to be remade. It could be because of a miscount in despatch or the product has been "lost" by sending it to the wrong place. The unplanned changeover KPI is a good indicator of the ability of the factory to get it "right first time".

Labour efficiency %

Labour efficiency is a major KPI for the food industry. Labour cost can be up to 30 or 40% of the overall cost of the product and the cost can easily get out of control if not monitored and managed. Some factories are highly automated and in those situations the labour efficiency becomes less important.

The simple form of this KPI is stated as: "How much did I spend on labour compared to how much I should have spent?" For example, at the end of a shift, it is possible to know how many labour hours you have used, by simply adding up the number of people and multiplying by the length of time they were paid for. It is also possible to calculate the expected labour hours that should have been used to make the food that was produced. Comparison of these two numbers will give you a labour efficiency percentage. By careful measurement and keeping a track of the KPI, the impact of decisions in the factory can be seen and improvements in the KPI can be made.

The factory decisions should be focused on the control levers that you have available. Careful management of labour in the factory will allow the cost to be minimised. The aim of this KPI is to try and improve the quantity of valuable work done by each person and to minimise time where the labour is paid but no valuable work is done. Lean Manufacturing techniques can identify areas for improvement in this vital area.

Machine efficiency %

Machine efficiency is simply stated as "how much did the machine make compared with how much it should have made in the same time." Everything about the usefulness of KPIs applies to this one. This KPI takes on a very important role in factories that are automated. The output of the machines is an indication of just how good the factory is at making its products. Machine efficiency % KPI is so important that it is often replaced with its more sophisticated cousin, OEE.

By now you will have an idea of how KPIs can be set up to provide information on what is going on in a food factory. Factories are very busy places and sometimes it can be difficult to keep an eye on the most important things. What is important will change from time to time, so the KPIs used will change. Do not fall into the trap of making too many switches. It is important that the core KPIs are consistently monitored with others introduced just before and during an improvement project, to provide information on the results of that work.

It is usual for a food manufacturing business to run with five or six core KPIs, and then one or two that are there for a shorter period.

After looking at KPIs for people performance and for machinery performance, here are some looking at the product itself.

Questions

- The area of machine KPIs normally is full of different measures.
- Have you got any not listed above?

Product KPIs

This group of KPIs focuses on the product being manufactured and will allow the regular measurement of performance in this area.

Average giveaway of packs produced

In packed foods, the weight of the product has to be tightly controlled. A big slice of the overall cost is the material that makes up the food. If products are heavier than required, this excess weight is sometimes called "giveaway". This is food that is manufactured and packaged and effectively given away to the consumer because of a lack of control somewhere in the production process. A KPI that is focused on giveaway will allow the management and the teams to monitor their performance in this area.

% of packs with faulty seals

One of the worse things that can happen in a food factory, from a cost point of view, is to get your product all the way through the process and still not be able to send it out to your customer because the seals are faulty. Measuring and monitoring the percentage of products that are rejected for this reason will help improve the performance of the packing operation.

Number of packs reworked

Rework is where product and materials end up, for some reason, having to go through one of the process steps more than once. For example, the product that is rejected for faulty seals may end up being passed back to the packing machine in feed to be rewrapped. A KPI here will let you know the ability of the wrapping system to get it right first time.

Average quality score

Your factory will have a method of scoring the quality of the products. The average quality score will give a view of the balance that is struck in the factory and the attention to detail required in making that product to specification.

Number of technical non-conformances per shift

Most food factories employ a team of technical auditors who check that all the systems and services in the factory are operating to specification. Where a failure is found, a technical non-conformance is often raised. A KPI that monitors the number of non-conformances will help the teams to keep an eye on their ability to follow the procedures correctly. A gradual increase in the number would indicate that there was a drift in the performance and there may be need of some corrective action such as retraining.

Number of packs rejected

The number of packs rejected allows a simple way of monitoring the ability of the manufacturing system to get it right first time.

Total weight of waste during the shift

This may be best expressed as a percentage of the total output, to make sure it is clear when the team had a good day and also when they had a bad day.

Daily hygiene audit score

Most food factories will conduct a daily hygiene audit to ensure that this important part of their work is measured, monitored and rapid action taken if the scores start to drop.

Bacteria counts on finished product

One of the effects of poor hygiene in the factory can be an increase in the numbers of bacteria in the final product. This is another way of measuring performance.

Questions

Again, product KPIs are popular as they start to be created by your quality management system and data capture through the QA teams:

• What are your KPIs that are focused on the products that you manufacture?

Process KPIs

Average temperature of materials arriving on the line

Here is an example of a KPI that could be implemented to help address a process problem. If the refrigeration of the materials arriving on the line is suspected as causing problems, make it a KPI for a few weeks to enable it to be measured. Remember that this type of KPI is there for a specific short-term period and should be removed again as soon as the issue is proved to be back under control. Unless this type of KPI is removed, it will lead to everything becoming a KPI. The administration and effort required to collect and process the data would then be too great. The production area would be swamped in a sea of KPI data.

Percentage of process check failures

It is a good idea to carry out a formal audit of your process on a regular basis to check that all is well. If the oven says the baking time is 22 minutes, what is the actual baking time? If the automatic mixing system says it has put 204 kg of flour into the mix, what is the actual weight? If the freezer says it is running at minus 18°C, is it really at that temperature? Process checks need to be carried out in a scientific way to make sure that the check is reliable. A KPI could be required at some stage if it is felt that process controls are not reliable and that the products are suffering as a result. A more direct example of a process check is a calibration routine. The weighing machine in your factory will need to be calibrated on a regular basis. The temperature and pressure gauges also need to be checked. Even the temperature indicators on sealing machines are worth calibrating, to ensure that you are working with accurate information (see Figure 4.5).

Questions

Process KPIs are popular in automated factories, as the data can sometimes be captured automatically:

• Have you got any examples of this?

Finally, here are some KPIs that might be used in your factory to help control and motivate people in the area of health and safety. I am sure you can see the relevance of regular monitoring in the areas below. As with all KPIs, the details and rules used to calculate the number will be tailored to suit the conditions in your particular factory. It will later be

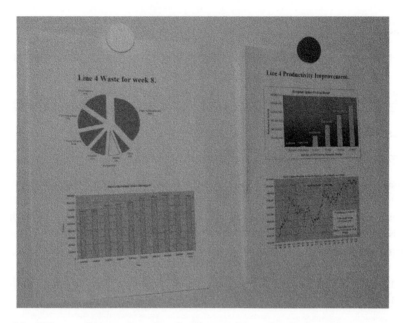

Figure 4.5   KPIs are often monitored so that the results can be displayed for everyone to see. This acts as a motivational poke in the ribs, as well as a pat on the back.

shown that some KPIs can be used to benchmark the performance in your factory with that of another. In this case, care has to be taken to ensure that you are comparing apples with apples. The definition of the KPI has to be precise and well understood if benchmarking between businesses is to be undertaken.

Safety KPIs

- Percentage of staff task rotated during the shift
- Number of accidents
- Number of near misses
- Daily safety audit score

Questions

KPIs in safety are popular:

- Do you measure days since the last lost time accident?
- What else do you do here?
- Try to think of the rules that could be applied in your business, to come up with a useful KPI for the areas above.

These types of measure will be looked at in detail later in this book and calculations will be made to come up with the number required.

Each of these KPI will take resources and effort to control, so it is important to select just the KPIs that are important indicators to your overall performance. For example, a rising average pack weight KPI would indicate a lack of control or training in this vital area; that lack of control or training will also show itself in other areas, such as a daily hygiene audit score or an increasing level of accidents and near misses.

In the land where everything is a priority – nothing is a priority.

It is important to decide where to focus effort and resources. The complexity of the food industry, with its competing demands, makes the selection of KPI a little trickier than other industries, but it is possible to have a high level of control without overburdening the business with unnecessary paperwork.

Communication of KPI information

The trick is to divide the controls into sections and ensure that individuals are clear about which sections of the overall picture they are responsible for monitoring. When tasks are divided like this, communication and teamwork become vital to ensure that the business optimises its performance.

The communication and teamwork is an issue among all groups of people, but its importance increases when the team is split into different departments, different shifts and even different factories or sites. The food industry is so large that the businesses within it have used many techniques to grow and these often lead to seven days a week working patterns, multi-shifts and supply chain systems, which mean that issues on your production line have been a result of the action of another team elsewhere in the business.

It would be easy to sit back and say: "not my fault" or "nothing I can do about it", but for performance to improve it is important that the parts of the bigger team are involved in some way in the performance of your department, your production line, and your machine. This can be achieved by using performance KPIs to feed back to the whole team, not just the local one. It is important that KPIs are measured, recorded and communicated.

How can this communication be *made* to happen?

Communication with the local team

This is the team on your shift, working alongside you.

Communication with this team should be simple; they are all here, they all talk with each other, they all are focused on getting the job done. While this situation is correct, it is important to remember that KPIs are all about measuring performance and taking action on the causes of poor performance. This requires the team to work closely together, as it is a situation where things are going to be changing in small ways. A change made in isolation from the rest of the team could cause unforeseen effects that could actually make the situation worse.

Case Study

The line stopped yesterday because we ran out of grated cheese, because there was a breakdown on the grater.

Corrective action is to get further in front with the grating operation to prevent running out. On the face of it, this would work and the person running the grating machine may see this as a good solution. However, there are some unexpected effects of this plan that the larger team could have pointed out.

We ran out of cheese yesterday – what can we do to prevent that happening again?

Technical supervisor – The issue of holding larger stocks of grated cheese is that the cheese will clump together with time and then this will cause a quality fault in the final product.
Hygiene supervisor – The issue of holding larger stocks of grated cheese is the increased number of tubs required. These are available but it would starve the factory of tubs that are used in many areas.
Stores supervisor – The issue of holding larger stocks of grated cheese is that the space required to store grated cheese is about twice that of the cheese in a block. That will put pressure on storage space. Stock rotation in the store will become more difficult if the store is fuller.
Looking at the KPI records for the line – the number of times the line has stopped waiting for cheese has been three in the last month.
Total lost time is 16 minutes. The reasons for the stops have been twice because of a change of plan, the wrong cheese had been grated, and once for the grater breakdown. The team decide that to increase stocks of grated cheese is unnecessary at this stage and that the situation of grater breakdowns will be closely monitored over the coming weeks to see if any change is necessary.

This example shows that KPIs can be used to assist in the management of complex situations, by focusing the whole team on the issues. As a result, decision-making by the team can be improved, because they can make decisions based on facts and records rather than just reacting to a recent issue or having to make a decision based on "gut feeling".

Short interval controls

Short interval controls (SICs) are the tools of managing a fast moving situation, by having up-to-the-minute information to prompt action to be taken and to keep the process running optimally.

An SIC that is away from its target should indicate that action needs to be taken quickly rather than waiting until the end of the shift to discover that a problem was occurring.

Examples of SICs in the food industry would be:

Machine run rate

How many good packs have been made in the last 15 minutes. Is that what is required to meet targets (see Figure 4.6)?

Figure 4.6   The speed control may show you the instantaneous speed but what has the output been in the last 15 minutes? Is that enough to meet the plan (picture courtesy of Ishida Europe)?

Average weight

What is the weight of product that has been given away in the packs produced, because they are overweight, in the last hour of production (see Figure 4.7)?

Adherence to plan

Is the line running the planned product at the planned rate? Is the line on schedule? Did the last changeover take eight minutes, like the plan said it should? Adherence to plan is a very powerful SIC. If a production line drifts away from the plan, it can have massive effects in the business. The management of those effects can be very costly. A production line running behind plan will have implications downstream in the packing and despatch areas. There will be delays in their work that will knock on into the distribution operation, with delays to vehicles and even late deliveries to customers. A line that is behind plan will also have implications upstream. The line will not be using materials at the required rate and that could cause service departments to slow down or stop as storage space gets used up in the middle of the factory.

The effects of being behind plan are reversed if the line is ahead of plan. Despatch gets full as vehicles cannot deliver early to customers and the service departments have to speed up to keep pace. In speeding up, it is possible that corners are cut or mistakes are made, but certainly the efficiency of the operation as a whole will be less that optimal. Running at the planned speed on the product will allow the factory to perform at its designed speed. If it becomes apparent that parts of the factory can run at a higher speed, it may not be more efficient for the factory as a whole to allow them to do that. The factory has to remain in balance, if efficiency is to be maintained.

Figure 4.7   The information will be there on your check weighing system. The giveaway for the last hour is a very good SIC, but you still need to take corrective action. You still need to pull the correct levers to get the situation back under control (picture courtesy of Ishida Europe).

Headcount

Is the correct number of people working on the production line? If more people are working on a production line than were planned or are needed, then the line will be costing more than it should. In the food manufacturing industry, profit margins are small and the business may not be able to afford that extra cost for an extended period. The need to get the crew size down to the planned number is vital if the planned profit margin is to be achieved. There will be instances where, for a short period, extra crew may be required. This could be because of an equipment malfunction or a raw material problem. As soon as the underlying problem is solved, the additional crew needs to be removed from the line so that it operates with its planned headcount (see Figure 4.8).

SICs are vital to the management of food production operations and run alongside KPIs in providing information for the people working on the line. SICs provide control information for instant actions to return the line to its planned performance. KPIs are part of a medium-term control where the performance of a whole shift can be reviewed and decisions can be taken by the team.

Communication of KPIs

The information contained in KPIs and SICs is very powerful and this power can be released into your business, providing people know about it. The communication of KPI and SIC information is vital if performance is to be improved. The use of start of shift

Figure 4.8   Imagine being in control of this department. Is the headcount correct at the moment? Has John come back from the toilet yet? Has Elaine returned from the wages department? How would you ensure that there is exactly the correct number of people here? Too many and its costing too much, too few and the production plan might not be met.

How does this team know that they are on schedule to meet the delivery expectations of the customer? How is the yield today following the problem yesterday? SICs would provide the answers.

Questions

- What SICs do you have in your factory?
- How do you know if you are having a shift that is on plan in respect of time, labour and materials usage?

---

meetings to keep people informed of KPI information will be looked at later, but first SICs themselves will be considered.

SICs need to be communicated live to the people who need the information. In some factories this is done by word of mouth or the ability for everyone to see the information in an easily digested form, say a graph. Some factories invest in large displays mounted over the line to carry vital SIC information to the people in the area. This is often linked to automatic data collection from a checkweigher or production machine.

KPI data is a great way of measuring performance of a machine or shift and the information can be used to improve performance overall. Some of the ways of using the information are:

Comparison

It is possible to compare KPI data with the data from last week, last month, and last year, and as a result, a view can be taken if performance is improving. Comparisons can also be made between shifts when running the same process, so red shift consistently get a 2% better yield than blue shift, breakdowns are 6% less on blue shift, quality failures are always 3% greater on a Monday, absenteeism is 5% worse in the weeks leading to a Bank Holiday.

Analysis and comparison of the KPI data does not solve the cause of the difference in performance, but it can point the way to better performance.

Sharing of KPI data

It is only when KPI data is shared with others that performance can be highlighted and improvements made.

Figure 4.9   Getting the "big team" to work together is a key factor in the success of Lean Manufacturing in the food manufacturing industry.

The difficulty of KPI data is that, if not correctly managed, it can cause unhealthy competition among the people being measured. A level of healthy competition is a good thing and can drive people to better performances, but it has to be recognised that, sometimes, the competition can go sour and people try to win the unofficial contest to be best without thinking about the bigger picture. They can be encouraged in this unhealthy competition by being congratulated for their outstanding performance, when in fact they performed well at the expense of another shift or department.

Overall the business can be worse off, with cost savings made by the high performing operation being outweighed by additional costs elsewhere. It is only by reviewing the whole performance that a balanced judgement can be made and it is only by working as a "big team" that company performance can be enhanced. The organisation of the "big team" and all of the little teams is important to ensure that a balance is met and company performance improves as well as individual and small team performance (see Figure 4.9).

For example, a team in the packing operation of a business has found a way of increasing the speed of its output significantly and as a result cost savings of £2000 per week are made in the labour costs of their area. Unknown to this team, their increased output is putting additional strain on the warehouse area, which is not equipped to handle the extra output. They have to take on additional crew to be able to get this product onto pallets at a cost of £600 per week. The other knock-on effect has been an increased rejection rate for the products being packed. Waste has gone up by £1500 per week. The saving made in the packing area has resulted in a net loss of £100 for the company. Now, obviously, there will be improvement activity in the area of rejects and pallet stacking, so the savings

Figure 4.10   Thinking about the whole supply chain has allowed big savings to be made in one area, with costs only slightly increasing in others (picture courtesy of Ishida Europe).

may reappear at a later date, but a coordinated approach between the teams would allow the improvements to be introduced with no period where losses occur.

Supply chain implications for KPIs

Some companies have taken this "big team" concept to its logical conclusion and included suppliers and customers within the team. That way, improvement in the whole supply chain can be made. For example, waste has been greatly reduced in some food retailers by placing orders for their requirements later and demanding a faster and more flexible response from the manufacturer. This has caused issues in the factory as inefficiencies creep in because of smaller batch sizes and shorter runs. This leads to more changeovers and so even more inefficiency. Overall, the supply chain efficiency has increased and as a result retail prices can be reduced to generate more volume. The extra volumes generated then feed into better factory efficiency and also better raw material and packaging material prices, as the orders for materials increase (see Figure 4.10).

Monitoring KPIs

Once KPIs have been chosen and targets set, the next step is to develop the system to monitor the performance and make sure that the KPIs remain at the front of people's minds in the business.

Calculation of KPIs

A KPI is most useful if it is a measure that is under the control of the people in the factory, rather than one that is influenced from outside their area.

A KPI of the total quantity of product sold by the company on that day would be of interest, but the real information and decisions are taken around the quantity of product made, not sold. In some factories making short shelf-life products, the quantity made will usually be the same as the quantity sold.

Another KPI could be the quantity of labour used in a day. The number of people working or the number of man hours worked would be useful information.

Looking again at the quantity of product made, this number is influenced by many factors, not least of all the quantity of products sold. Food sales are seasonal and are influenced by the day of the week, the weather and even the promotional activity. All of these factors are outside the control of the production team and so a measure of products made is generally not a good one to monitor the performance of the team.

The real power of KPIs is when both of the items above are calculated into one KPI. If the quantity made in a day was divided by the number of man hours used that day, it would produce a single number that describes both to aid decision-making.

The products made would be of good quality, man-hours worked would be acceptable, and products per man hour are a great KPI. The single number measures how efficiently the products were made and is not influenced by the number of people in the factory or the quantity of products made. A KPI of products per man hour can be used as a performance measure and can be used to compare the performances from day to day, shift to shift, product to product and even factory to factory (see Figure 4.11).

The creation of KPIs using two pieces of data can be used in many ways:

- Waste measured in kg is alright, but waste per product would be better.
- Number of lightweights is enough, but lightweights per 100 products would be better.
- Number of accidents is acceptable, but accidents per 1000 man hours are better.

Questions

Think about the KPIs that you currently have to help control your factory:

- Make a note of them here.
- How many of them are calculated using more than one data input?

After calculating the usefulness of KPIs, ways need to be considered in how to monitor them. The monitoring of KPIs can take several forms, so here are a couple to think about:

Figure 4.11 The output of this manual line would be interesting to know but the output per man hour could be used as a measure to start improvement activity and would be an ideal KPI for this factory.

Daily/Weekly Operating Report (DWOR)

The DWOR captures the KPI information and records it in a document, which is usually a spreadsheet with daily results and a weekly total to date, hence DWOR. The DWOR becomes the record of the week and can be used to monitor performance against the targets for each of the KPIs used. Some companies have a DWOR system that is computer based and as a result the spreadsheet can be programmed to be active, so the calculation of the KPI is automatically based on simple data input. The DWOR will then be one page that can be printed out for display in the factory and also be carried into team meetings where issues can be discussed.

Often on a DWOR, the areas where a target has not been met will be highlighted in red and areas where targets are met in green; this way it becomes apparent to everyone, even when simply walking past a notice board, of the number and position of reds and greens on the DWOR. This, by itself, will not motivate people to improve but is a useful prompt for people to get them thinking.

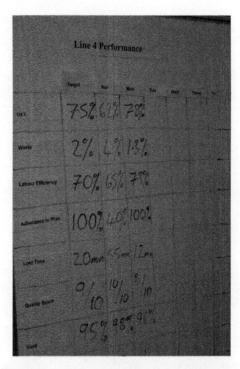

Figure 4.12 A KPI display can be simple to do. Wipe clean boards are very popular in food factories, with the numbers filled in by the person who is responsible for the KPI. Targets are always shown with numbers below target emphasised. I think that Line 4 here had a bad time on Sunday!

See Appendix 3 for an example of a DWOR for a food factory. That shows a data collection sheet with important features such as a three-shift system, so that performance comparisons can be made for all operations. There is also a target section. The DWOR becomes the place where people look for improvement opportunities (see Figure 4.12).

The concept of colour prompts is a big area that will be covered elsewhere in this book, but one important recent development has been a "dashboard" for a business.

A dashboard is a visual display of important business measurements, KPIs, rather like the dashboard in a car. In order to drive a car, you need information on speed, revs, fuel level, temperature, etc. This information helps you drive the car and arrive at your destination on time. In the same way, a manufacturing dashboard provides information that allows your process to deliver what is required. Health Warning – if you tried to drive a car simply by the information contained on the dashboard, you would not arrive safely; you would most likely crash into the first lamp post that you encountered on your journey. In the same way, trying to drive your manufacturing process simply by looking at the KPI dashboard is not sufficient and a crash would occur quickly (see Figure 4.13).

KPIs, DWORS and dashboards are important tools in a modern factory but they are not the complete answer to running an efficient business. They provide the foundation of accurate and timely information by which the business is managed, but the business still has to be managed.

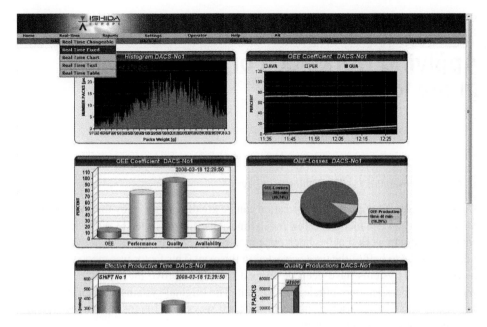

Figure 4.13   A simple dashboard to monitor OEE performance KPIs in the business (picture courtesy of Ishida Europe).

Summary – Starting to measure and quantify performance

This chapter started by thinking about ways to measure and record the performance of a manufacturing operation. This can be a complex task and it is important that the measurements are reliable so that they can be used as a basis of decision-making in the business.

Getting your hands on the levers was introduced in this chapter, which introduced the ideas of finding the factors in your business that really control its performance. It is these levers that need to be pulled and positioned in order to optimise what is going on.

Measuring performance is the vital first step in improving performance.

Chapter 5

Applying workplace organisation in the food industry

A place for everything and everything in its place.

The performance of a production operation can be greatly improved by the application of some simple rules and methods. How much time and effort is wasted by either looking for things, or redoing work that has been done incorrectly. Studies have shown that, in some workplaces, up to half of the time is wasted in this way. This factor was recognised some years ago and systems have been developed to help improve the situation and make people and processes more efficient as a result. The original systems have changed over the years and show themselves in many different ways in different businesses, but the basic principles are always present.

The system called "5S" is an example but there is no need to follow these rules rigidly. The secret is to apply the principles as they suit your business.

5S system of getting organised

The 5S system was developed in Japan and originally the names for each were, unsurprisingly, in Japanese. The names have now been adjusted to English and some of the original meaning has been lost in translation, but the basic principals still apply: Sort, Sweep, Standardise, Set Locations and Sustain.

The first S of the 5 is for "Sort"

This step of workplace organisation is to sort out what is required to be in the workplace and to take away unnecessary clutter so that only essential items are present.

This sorting process is very important in the food industry, as it keeps the workplace uncluttered and easier to maintain from a hygiene point of view. The sorting process will also remove potential foreign body contaminants from the area and as a result reduce the risk to the food being manufactured.

The kinds of clutter commonly found in food production areas are:

- Sets of change parts for products that are no longer manufactured
- Spare machines that are never used

Handbook of Lean Manufacturing in the Food Industry, First Edition. Michael Dudbridge.
© 2011 Blackwell Publishing Ltd. Published 2011 by Blackwell Publishing Ltd.

- Last week's production plan
- Spare pens
- Broken bits of machine
- Scrapers and hand tools
- Raw materials from a new product trial
- Keys for padlocks that no longer exist
- Spare tickets and labels
- Redundant packaging.

…the list goes on.

Questions

Before you go on, you should have a look in your production area and list the items that are there and do not need to be. While you are looking, ask yourself the question:

- Why is this stuff here?
- Why do spaces fill with clutter, unless it is managed and controlled in some way?
- The final question is how can you stop it from building up again?

How do we decide what is required and what is unnecessary?

This could be quite a simple decision but care needs to be taken not to remove something that is required by someone. There is nothing more annoying than not being able to find that item "that you knew was there". There is a method of carrying out this selection of things to remove and this is called red tagging.

The item considered for removal has a red tag attached to it, giving the name of the person planning the removal and the date when removal will occur. Obviously if an item that you use suddenly has a red tag attached, then it is a prompt to see the person named and justify why the item should stay. This is very important when the production operation is spread across several shifts. What is seen as unwanted clutter by one team of people may be essential to the operation of the other shift. Perhaps the spanner you are thinking of removing is essential to the night hygiene crew. Red tagging should be applied with a degree of common sense. Things that are obviously rubbish or unusable can be moved straight away. It is when more substantial items are considered, that a red tag should be applied. The aim of red tagging is to gain a level of control over what is in an area, without removing items that are needed. Red tagging gives everyone an opportunity to come to an agreed decision, rather than the decision being made by one person. If there is an agreed removal the item will, or should, stay removed. An enforced removal will usually cause the item to reappear soon afterwards.

For example, a red tagging exercise in an engineering store, by removing the items that are not required, will make it quicker and easier to find the needed items. The racks will also be easier to clean, easier to stock check and become a more efficient use of the valuable space.

Removed but still available

The aim of the red tagging exercise is to identify items that are needed in the production area but it also indicates items that are required but on a less frequent basis. It is good practice to set up a system of storage for items used in the business, where items used frequently are stored in the immediate area of their use. A scraper that is used to clean the surface of a conveyor every hour is stored in a marked location near the conveyor. A pen that is used every 15 minutes to record pack weight is stored next to the clipboard that contains the paperwork, and both the pen and the paperwork should be next to the weighing machine used to make the checks. In this way, workstations can be established in the area that contains all that is required to carry out that task. Workstation and storage areas need to be well marked in a way that makes it very obvious when an item is missing. A good example of this approach is a shadow board for the storage of tools. Instead of a toolbox, storage is achieved using a wall-mounted board. The shape of each item is marked on the board, so if the item is missing it is immediately obvious (see Figures 5.1, 5.2 and 5.3).

Items used less frequently are stored close to their point of use for easy access but not so close that they cause congestion in the area. A change part for a wrapping machine that is required once or twice in a shift should be stored near to the wrapping machine to facilitate quick changeovers when required. The tools, spanners, etc. required to make the changeover should be at the same location. It is possible to make a mobile storage location. A trolley containing change parts and the tools required for a changeover is designed like a shadow board, so that it is easy to see if anything is missing. When required for a changeover, the trolley is taken to the line to make the access to tools and parts easy and to speed up the changeover process. This kind of system will help reduce the downtime on the line and stop you hearing people shout: "…has anyone seen the 15 mm spanner?"

Items used infrequently should be stored out of the production area. A set of calibration weights for checking weighing machines once a week and a box containing holiday request forms should not be in the production area, except when they are being used, and they should then be removed again immediately afterwards. The storage of these items still has to be controlled to ensure that the item is always in the same location. If this is done successfully, there will be no need for everyone to have a few holiday request forms in their locker, as they will know where to get one when they need it. The calibration weights will be stored in a marked location so that all the shifts know where they are. The engineers know where they are, the Quality Assurance Team know where they are and even the person from the local trading standards department knows where they are.

Finally, items that are never used should be disposed of and not kept for a rainy day when, "I am sure we will use it one day – it is too good to throw away."

Somewhere at the back of every factory is an equipment graveyard; this is sometimes necessary but rarely well managed and is often full of equipment that does not work. Sometimes a similar picture can be seen inside of the factory too, with unused equipment and machines causing clutter, taking up valuable space and requiring hygiene time to prevent it becoming a risk. Lean Manufacturing cannot afford the time and effort that will be spent looking after equipment that is not required. By removing the equipment, you are reducing cost. By setting storage locations and workstations and by using shadow boards, cost is reduced. Twenty seconds looking through a toolbox for the correct size of spanner

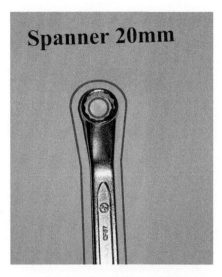

Figure 5.1 A shadow board used to store tools. These can also be created to store machine change parts or even clipboards for quality records. A shadow board can even be created for a single item that is required to be available in a particular location.

Figure 5.2 A shadow board used to store tools. These can also be created to store machine change parts or even clipboards for quality records. A shadow board can even be created for a single item that is required to be available in a particular location.

might not appear to be a lot of time to waste. But if you are looking each time there is a changeover or you need to make an adjustment and you do 6 changeovers or adjustments per shift, and there are 3 shifts per day, that 20 seconds adds up to 36 hours of lost time per year; that is a week of someone's time. Imagine what the effect is if the spanner goes missing from the toolbox (see Figure 5.4).

One of the radical steps that is sometimes taken to prevent a build-up of clutter is the removal of storage space and locations.

Figure 5.3   A shadow board used to store tools. These can also be created to store machine change parts or even clipboards for quality records. A shadow board can even be created for a single item that is required to be available in a particular location.

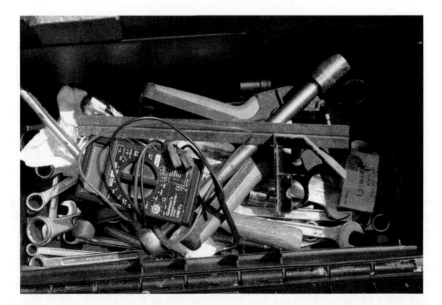

Figure 5.4   "Hey John, have you seen the 20 mm spanner? I've looked twice and I can't find it." "Let me have a look," says John, "I'm sure I saw it earlier."
Both people are then looking for the spanner. Meanwhile the line is stopped and the others working on the line stand and wait for the changeover to be complete. Add up the lost time now. What would be the effect if the whole toolbox went missing?

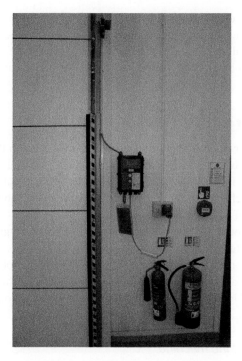

Figure 5.5 The flat top on this electronics cabinet means it can be used as an unofficial storage location.

Storage location removal

By reducing the number of places where things can be stored makes the task a lot easier. Some of the techniques used are:

- **Replacement of all toolboxes with shadow boards** – Toolboxes are notorious places for clutter and they pose a contamination risk but also an efficiency risk. A cluttered toolbox makes it more difficult to find the spanner that you need and this might have an impact on a changeover time.
- **All flat surfaces are sloping**. The tops of machines are made to slope to prevent these surfaces becoming a storage location. It is no longer an option to store a spanner on top of a machine. It must be stored correctly on the shadow board provided (see Figure 5.5).
- **Open mesh storage lockers** – Where items are required to be securely stored, they are placed into wire mesh lockers rather than solid ones. This has the effect ensuring that no unnecessary items find their way into the same location. The contents of the locker are constantly on view, so discipline can be maintained.

  A wire mesh locker does not prevent clutter, only people can do that, but it does make the clutter very obvious and therefore easily managed. This design of locker is very good for the storage of clean-as-you-go equipment close to the production area, such as brushes and squeegees and maybe a few spare black bags for waste disposal.

Now the area is sorted and all unnecessary items have been removed and the necessary items have their own storage location. Already your factory will be starting to look a bit different and people will have noticed that changes are being made.

The production area looks tidy and organised, the first step towards efficient, low cost manufacture.

The second step of the 5Ss is "Sweep"

Once the sorting has been done and only essential items remain in the production areas, the next step is called Sweep. In a food factory, this step is essential from a hygiene point of view. Food hygiene requires that machines and equipment are clean, but Sweep goes one step further. If machines and equipment can be returned to "showroom condition", it provides a great deal of advantages to the business.

Ideally machines should be clean both inside and out. Normal hygiene will concentrate of the food contact surfaces Sweep also improves the conditions inside of the machine on drive pulleys, motor housings, pneumatic systems, electrical control panels.

Safety first – before any access is gained to these parts of a machine, it is vital that the machine is isolated and locked off to prevent safety issues for the person doing the sweeping.

The advantages of machines kept clean in this way are many.

Regular inspection of the machine

The machine covers are removed on a regular basis and inspection of normally inaccessible areas is made possible. If the person doing the deep clean is correctly trained, then it is possible to incorporate this clean as part of the regular preventative maintenance system. While the pulley belts are cleaned, they are also inspected for wear and tension problems. While the chain and sprocket are cleaned, they are also checked and lubricated.

Quicker breakdown recovery

Should a machine break down, a clean machine is easier to work on for the engineer fitting the new part. Tension rollers move, bolts can be undone, door seals are not damaged. Imagine the difficulty of spotting an oil leak on a dirty gearbox, compared to the same fault on a clean one.

A deeper understanding of the machine

The sweep clean would usually be carried out by the machine operator. The operator's knowledge of the machine will improve and as a result the machine will be run more effectively.

Improved reliability

A clean machine will run more consistently and produce less rejects.

When a large quantity of sweep cleaning is occurring in your factory, it will not be long before you realise that a lot of time and effort is be spent on returning the machine to as new condition on a regular basis. You will have seen the improvement in performance, but the cost of that performance increase could be high. One area of continuous improvement is to try and reduce the workload by working smarter rather than working harder. It will not take

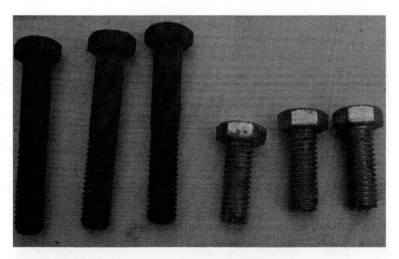

Figure 5.6 The replacement of long bolts fixing machine covers with shorter bolts will speed up cover removal for the regular cleaning. First check that the short bolts will be acceptable to the engineer in the improvement team.

long for someone to come up with the idea that if you can stop the machines getting so dirty in the first place, the sweep clean could be carried out more quickly. Ideas for removable catch trays, covers for electric motors, and even making holes so the debris falls out of a machine rather than staying inside, will soon be suggested by the team. There are also techniques that allow easier, but safe, access to the inside part of a machine, so less time is taken in the removal of the guarding and covers ready for the sweep clean to occur. These techniques return the machine to better than showroom condition. The team will have modified the machine to fit better with how the machine is used (see Figure 5.6).

Once Sorted and Swept has been carried out, the production area will be starting to look and feel a bit more organised. This stage of the 5S process is critical to the success of this method of workplace organisation; it is vital that the hard work completed is not wasted. The workplace needs to stay sorted and swept and there will need to be a new way of working among the team. Items must be returned to their correct storage location, deep cleaning must continue, and people need to maintain the disciplines that are required to keep the production area in tip-top condition. In the food industry, change is a way of life and there will be new materials and machines introduced as well as materials and machines becoming redundant. There will be new people starting who will need to be shown the new way of working to ensure they contribute to maintaining the standards.

The next of the 5Ss is "Standardise"

This is where the methods used in carrying out any task are the same, no matter who is carrying out the task. A standard method for every task. This is a big job, but is the best way of improving efficiency, reducing mistakes and keeping the quality of the output to the level required.

To start the process of standardised tasks, it is a good idea to put a group together that has that task as part of their role. They can work together to come up with a

Figure 5.7   The replacement of standard bolts with ones that can be undone without the use of a tool might speed things up a little. Safety is a high priority, so this should only be used in areas that are not safety critical.

method that they can agree on as a standard method. While they are working on this, they could also discuss ways in which the task could be improved and try and use the best method that they can think of in terms of safety, efficiency, quality and cost.

The new standard method needs to be recorded in some way, so that others can learn the method and also carry out the task correctly. It is good practice to record the method in writing and also use photographs or video clips to ensure that the method is fully documented (see Figure 5.8).

As there are hundreds of tasks in most food production areas, it is difficult to decide which one should be carried out first? Probably the best way of taking this forward is to start with a list of all tasks that are to be covered. Tasks around a tray sealing machine will include:

- Pre-shift start-up checks
- Starting the machine
- Changing the date code
- Alignment of date code in target box
- Changing the roll of film
- Testing for seal strength
- Testing for leakers
- Changeover of sealing heads
- Cleaning of sealing heads.

Any and all tasks need to be on the list. It is then a matter of deciding which ones to tackle first, putting a small team together to look at the task in detail and come up with a standard method. Remember, because this is a small team, it is possible to have several of these standardisation projects running at the same time.

Figure 5.8 This is a "before" 5s picture of waste segregated ready for recycling. The standard method is created once the area has been sorted and swept. This is an interesting area, because mess is constantly arriving and the task undertaken is in controlling that mess. I am sure you have some suggestions here for the standard method to address the issues of safety, efficiency, quality and cost.

With more manual work, the list of tasks can be just as long. Tasks around the weighing of dried ingredients will include:

- Use of dust masks
- Opening of bags and sacks
- Segregation of allergens
- Stock rotation and tracability of ingredients
- Calibration of scales
- Use of scales
- Colour codes for scoops.

Once again, any and all tasks need to be standardised.

Questions

- Does your factory have a scheme of SOPs?
- Could there be more use of this type of system to improve performance, by finding the best way to carry out a task and then making sure everyone does it that way?

Once the SOPs have been agreed and recorded, they become the basis of your training system for the people carrying out the task. They are also subjected to regular checks by the Quality Assurance Auditors, who will make sure that the standard methods are followed.

The other advantage of having a SOP is that it forms the starting point of improvement activity. Improvements will be made to these methods and procedures, which will then be re-documented and retraining of all who need to know.

Figure 5.9 It can be seen here that the floor of the area has been marked to ensure that items are in their set location. It is like a shadow board on a grand scale. It becomes quickly apparent when an item is missing from its location. When items are stored in a random fashion, they are difficult to find and will go missing easily, having a knock-on effect in production efficiency (picture courtesy of Ishida Europe).

The fourth part of the 5s system is "Set Locations"

It is important for Lean Manufacturing techniques that the location of everything is set. This is from the storage location of a spanner used in a product changeover to the location of the clipboard used to collect waste data from the process as part of a project.

Imagine the amount of time spent in a factory looking for items. Individual productivity can be greatly improved by just knowing where the things you need are. Set locations for everything has big benefits for all in the business (see Figure 5.9).

Questions

Make a note of the last time your performance was slowed down because you could not find your car keys:

- Do you have a "storage location" in your home for important items such as your passport?
- How does your factory tackle the issue of storage locations in production areas?
- Could the storage of vital tools and equipment be improved? Write an example of one item that is always going missing.

Storage locations are best marked and labelled, so that it is clear when something is not in the correct location. Floor marking can be used to designate the location of a pallet of raw material, for example. Pallet racking can be labelled so that each pallet location has a number that can be used to find the material you are looking for in the warehouse. Shadow boards can be used for small items such as tools and change-parts.

Large machines are often bolted to the floor, so location is fixed but some of the ancillary machines around the main line may need to be in an exact position to work correctly. Hoppers depositing sauce into a tray will need to be carefully aligned to work correctly. Their location should be obvious to everyone.

Last of the 5s is "Sustain"

This "s" is all about keeping the systems in place, so that the factory does not backslide and performance suffer as a result. Typically, sustain is all about the training of people into the standards required. 5s must become the way that the factory is organised and the people in the factory need to abide by its rules if improved performance is going to keep happening. The pressures on the shop-floor to "just leave it here" or "miss that routine machine clean" are large and always there. It is the sign of a good factory where these pressures are not allowed to interfere with the 5s system and the factory is kept at a high standard the entire time.

5s becomes the way that the factory is operated and not an additional chore to be carried out. As a result of it being the only way that the work gets done, the training and development of new people in the factory is essential and is seen as key in order to sustain the system.

5s is a lot about attitudes and behaviours of the people working in the factory and those attitudes and behaviours must be universal if the system is to be sustained over a long period of time. Having 5s operating in a department will not be sustainable if the manager's office is disorganised.

Implementation of 5s is often attempted one department at a time. This can work, but more successful implementations can come about by having a factory-wide campaign. A 5s month with daily audits of the changes can be made to work because of the spirit of competition between teams to get sorted, swept and standardised.

Improved performance through workplace organisation is a relatively easy step to take. The factory becomes more efficient through the work done. By being more efficient, costs of the operation will reduce. 5s is a good system to use in the food industry, as a lot of the sections in 5s also have a benefit in terms of product quality and food safety (see Figure 5.10).

Visual systems to help performance

There is a huge amount of information that is present in every food factory. In order to run effectively, the factory is as much about the control of that information as it is about the control of the food itself. In order to stay in control of a process, it is important that the information is easily interpreted. This can be achieved with the use of colour coding and other techniques.

Information in a factory may be in the form of a pressure gauge, a digital readout from a temperature probe, a list of weights from a QA record sheet, a verbal instruction from your manager, a tag on a tub of ingredients, a machine control panel, etc. All of this

Figure 5.10   5s can also be applied to areas around the production area. This changing room could be changed to improve the efficiency of entry into the factory.

information needs to be understood and decisions made about what the information means. This is a complex area for people working in the factory and so there need to be methods developed to make this as simple as possible.

The Visual Factory

This is a factory where information is provided in visual form rather than just written form.

For example, a pressure gauge on a machine, instead of being a pointer to read a number to indicate the pressure, would it be possible to put a green zone onto the gauge to indicate when the pressure is correct? A red zone can be used to indicate under- or over-pressure. Instead of having to read the gauge, the operator simply needs to see that the pointer is in the green zone (see Figure 5.11).

For example, a pipeline carrying flour to a mixer in a bakery would occasionally become blocked, but this only became apparent when the mixer stopped, waiting for flour. A visual solution to this issue was to insert a see-through section into the pipeline, so that the flour could be seen flowing.

There are many examples of ways to make your factory more visual:

- **Photographs** – to illustrate a particular part of a task or to identify a storage location with a picture of the item that should be in that location.
- **Colour codes** – red brushes to be used on floors only, not on machines. Red tubs are to be used for meat products. Green scoops are for handling organic foods.
- **Floor markings** – to mark the storage location of equipment and materials, but also to designate walkways and areas where items should not be left. Commonly, areas around emergency exits are marked to ensure they are kept clear.
- **Coloured paperwork** – the use of paperwork to collect information on the factory floor is common. Food Safety critical records, which form part of the HACCP system,

Figure 5.11 These gauges have a green zone but also they have a red light to indicate out of control situations!

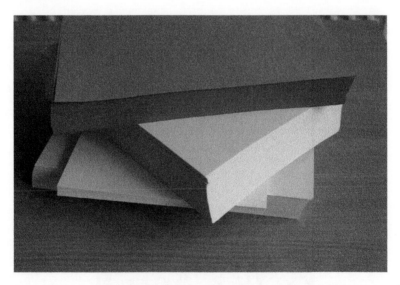

Figure 5.12 Different colour paper is often used for different areas of the factory or technical requirements.

could be colour coded pink to make them stand out on the shop-floor and emphasise their importance (see Figure 5.12).

- **Marked gauges** – this is as seen above.
- **Arrows** – to indicate a flow on a piece of pipe work is commonly done. How about arrows that have to be matched up when a cover is put back onto a machine after cleaning. It may make the refit a fraction quicker (see Figures 5.13 and 5.14).

Figure 5.13   Markers for pipelines make it easier to see what is where in a complex building such as a food factory. Easier means quicker, and quicker means cheaper.

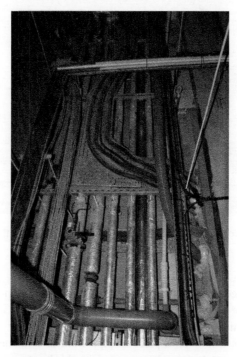

Figure 5.14   How easy is this lot to get right without the help of markings? A visual factory is an easier, safer and more efficient place to operate.

Figure 5.15   Notice the lack of a clock on the wall. It is vital in food production areas for there to be some way of telling the time. Not just so that everyone knows when tea break is coming up, but so that activities can be accurately coordinated. How often has someone turned up late for a team meeting in your factory because the clock in their section has stopped working.

- **Traffic signs for no entry, etc**. – these are very easy to interpret and can help in areas where the flow of people and materials can be improved.
- **Coloured hats and overalls** – these allow the easy recognition of team leaders and first-aid personnel on the factory floor, but can also be used to distinguish protective equipment used to handle meat or organic materials (see Figure 5.15).
- **Colour coded packaging** – the retail environment requires a range of food products to look similar and present a strong brand family on the supermarket shelves. This similarity can cause mistakes to be made in the packing department. Packaging can be differentiated by the use of small coloured patches in a convenient position on the pack. If well-designed, the coloured patch will be obvious on the magazine infeed to a packing machine, allowing a mistake to be spotted.
- **Visual standards for product and ingredient recognition** – it is difficult to describe what some food products should look like at various stages in their production. A photograph can be used to give a reference point for the operative. Visual standards can also be used to define the upper and lower limits for a product appearance (see Figure 5.16).

Standard Operation Procedures (SOPs)

The creation of an SOP, where one has not existed before, was discussed earlier. The difficulty with SOPs is keeping them up-to-date and in use. The SOP manual can sit on a shelf in the manager's office or the SOP can form part of the factory by being displayed and used.

Figure 5.16   Imagine trying to record what this meal should look like for your quality system. How about taking a picture of the acceptable product? It becomes a visual target for the team putting the meals together. It includes a lot of information about quantity and colour, position and surface texture. All in one picture. Visual standards of products are now commonplace in food factories (picture courtesy of Ishida Europe).

For example, the SOP for stock rotation and organisation in a storage area could be quite a wordy document or it could be built into the store itself with markings on the floor and a First in First Out (FIFO) system of stock management. That way, it is impossible for the SOP to be done wrongly without a great deal of effort. This technique of making the standard method the easiest method is the best solution to ensure that the SOP remains in use. It is the easiest method, so why would anyone want to do things differently.

For example, it is possible to achieve enforced FIFO using rollers. The new items are fed from the back and move forward as older items are picked from the front.

With some thought and reorganisation, it is possible to organise a factory to contain many the these easiest methods, but it must be recognised that this may not always be possible, especially where the factory is one built with flexibility in mind. The easiest operating method may be adequate for most products, but it may not suit every product. In these cases, people need to know the required procedure and follow it in every detail. That knowledge of SOPs can only come about through a system of training and competency testing, to ensure that people understand.

Training of SOPs

The SOP should be drawn up in a way that allows it to be used as a training document and people should be trained in the procedure with the document as a guide to learning in a logical way. Learning an SOP from someone who is currently doing the job is not acceptable without constant reference to the SOP document. Informal training, on the job, can be quick but will inevitably pass on bad habits and will miss out some of the knowledge required.

Training in SOPs is often linked to pay scales, with people able to increase their pay rate by becoming trained. The best schemes of this type will also include a check or test of the person being trained so that they can demonstrate their competence. These checks can then be repeated at intervals, to ensure that the standard methods continue to be used.

SOPs as a leveller of performance

Once an SOP is well established in a business and people from all shifts and departments fully understand the method to be used, it will by its nature level out performance, safety and quality differences between the shifts.

For example, the night-shift was really quick at carrying out a product changeover and as a result they had improved output because of the reduction in downtime. However, the method used by the night-shift did lead to product quality issues when the line restarted, and they also suffered a higher than average number of accidents due to the methods used. Clearly it would be great to get the high performance and not the quality and safety issues. In the real world that is unlikely, and it could be that the final methods settled on, as the SOP for that changeover, needs to have a slight compromise on performance in order to eliminate the safety and quality concerns. But once the SOP is installed, all shifts will be capable of meeting the same standard of performance, safety and quality. Overall, the business is better off in the long run.

SOP as a method of improvement

There are two factors here, the one mentioned above where a best/easiest method is developed and adopted in the business and this leads to improved overall performance. The second factor is as a result of developing and changing the SOP to be better/easier. Once an SOP is established, any improvement that is made has an impact across the whole business, rather than just for one individual or small team or shift. Therefore, any improvement in method will lead to an improvement for everyone. This is one of the key advantages of a robust SOP system.

For example, the task of weighing powdered ingredients into recipe batches for use in a cooking or mixing operation. The task occurs in several locations in the factory for different sections. One operative discovers a new method of opening and emptying bags that is both quicker and means that less powder remains in the bag when it is sent for recycling. Before an SOP was established for this task, it might be that the operative did not reveal the new method and as a result no one else could benefit. With an SOP in place, it is possible to make a modification, retrain the staff and quickly gain the full benefit across the whole business.

Think of the implications for this method of improvement if the factory concerned was part of a big group. An SOP improvement in another factory could help improve performance on your shift in your department.

Improving an SOP in this way does raise one point, which needs some thought. If the SOP is standard, how can someone try out different methods to see if they are better/easier? The answer is one where an individual should not play with the standard method; one of the key strengths of an SOP is that it is created by a small team of people. It is that small team who should work on the improvement, to ensure that there is no downside to the suggested change. In the same way that SOPs are created, they can also be improved and then the full benefit can be achieved by communication and training in the new method.

Summary – Applying workplace organisation in the food industry

Lean Manufacturing is all about getting organised and staying organised. This chapter has looked into a few techniques that are commonly used in food factories to improve their performance. The first of these techniques was called 5S and is all about workplace organisation. It is a five-step process to sort out the manufacturing and other areas, so that the task of manufacturing food can be the focus, and so the number of distractions to that task can be minimised.

The five steps are Sort, Sweep, Standardise, Set Locations and Sustain. A 5S programme will see efficiency, safety and product quality all improve and the success of a 5S programme can be measured by its impact on the KPIs of the factory.

Chapter 6

Improving flexibility and responsiveness

The food manufacturing industry is part of a complex supply chain that moves food from the farm gate to the dining table of the consumer. It is the demands of the consumer that drive a need for all parts of the chain to be flexible and responsive to changing patterns of demand. The food manufacturing industry is also very competitive, with an overcapacity in the industry and a tendency towards copycat or me-too products.

A food manufacturer may launch a unique product on a Monday morning, by Monday afternoon the product has been purchased by competitors and it will have been imitated by the end of the week. A couple of weeks to get packaging designed and delivered and suddenly the unique product has competitors alongside it on the shelves of the retailers. A food manufacturer cannot just use innovative and new products to gain a commercial advantage. Add to this the fact that competition will also exist in the form of pricing of products and it can be seen why, to be successful, the food manufacturer must be efficient, innovative, flexible *and* responsive.

Copycat food products are constantly launched to ensure that innovative products do not have long in the market before the competition arrives. A food manufacturer cannot simply rely on new products for improved performance.

The only response of a food manufacturer to all competitive activity is to build systems into its factories that allow it to be efficient and flexible and also to be able to respond to changing consumer trends and tastes. And all this in a business with very low profit margins and high performance requirements.

Case study

Manufacturers of pork pies are seeing a decline in their sales as consumers opt for more healthy options. In the past the manufacturers have invested in automation of the production of the pies in order to keep costs low. The company will have noticed the sales trend and embarked on reformulations of the original products, reducing fat and salt at the same time as developing new products in the healthier markets, which utilise the skills of the workforce and the machinery in the factory. Without this flexibility and responsiveness, the company sales would continue to decline and the company would ultimately close.

Handbook of Lean Manufacturing in the Food Industry, First Edition. Michael Dudbridge.
© 2011 Blackwell Publishing Ltd. Published 2011 by Blackwell Publishing Ltd.

How do companies organise themselves to improve their flexibility and responsiveness to the demands of the consumer?

There are many methods by which a manufacturing company can become more flexible and responsive. The first of these is not so much a physical change or investment in new machinery; it lies within the attitudes, behaviours and beliefs of the people within the company.

Customer focus

The first step towards a more flexible and responsive manufacturing company is to get everyone in the company focused on the needs of the customer. The customers and the consumers are, of course, two different sets of people. The consumers are the people who eat your food; they have needs but for the moment the focus should be on the needs of the customer; that is the next person in the supply chain. These will normally be food retailers or wholesalers.

It must be clear to everyone in the manufacturing company what the customer needs and demands.

This will include delivery of the right quantity of food – on time. A performance of 99% on time/in full delivery would not be seen as good enough for some of the major food retailers, so systems need to be in place to ensure 100% delivery on time/in full.

The customer will also demand that they place an order with the manufacturer as late as possible so that they can maximise the availability of your products in their stores without risking waste by having too much stock. Lack of availability in store is a key reason for consumers to switch loyalty and move to another retailer. If the consumer wants strawberries in February, then the store has to provide that choice or the consumer will look elsewhere. A retailer that runs out of milk or bread has a major problem, as 80% of people visiting the store will want those items in their basket. No bread or milk in the store will sometimes see consumers leaving half full trolleys in the aisles and going to the next retailer for their shopping. It is pressure on the retailers by consumers that cause them to set such high delivery expectations and require that they be met consistently (see Figure 6.1).

The customer will demand that prices to them are as low as possible to allow them to compete with other retailers. Consumers are very price sensitive in the area of food purchases and a retailer that fails to meet the value for money test will ultimately lose consumers.

It goes without saying that the customer will always demand that the quality of the product will be to the standard set and agreed.

The customer demands low cost, 100% service, quick response and unwavering quality, plus they want new products to excite the consumers along with imitations of innovative products from around the world and the ability to offer bargain deals to consumers in the form of Buy one Get one Free (BOGOF) promotions. How can a manufacturer meet these demands?

A balancing act

Meeting all these demands is not an easy task, but to do so while maintaining a healthy profit margin is, inevitably, a balancing act. Sometimes some of the demands could only be met at a great cost, so performance in this area of customer focus is rarely perfect.

Figure 6.1 On time/in full delivery requirements are needed by retailers to prevent out of stock situations on the shelves. On the other hand, too high a stock level would cause the retailer to have high wastage levels in store and incur large storage costs. A few years ago a major supermarket store would have a large stock room at the back to ensure that the shelves were always full. Modern superstores have very little space at the back of the store for stock. All available space is given to retail space.

Deliveries will sometimes be short or late. Quality will sometimes suffer slightly. It is the way that the manufacturer maintains this balance that will characterise the relationship it has with the customer. Large failures or a series of small failures could well indicate that a manufacturer is not focused on the customer and may well result in a loss of business.

Some techniques to maintain customer focus in food manufacturing businesses are:

Shift patterns and weekend working

The food industry is like any other, in that it needs to maximise the output of the factory and machinery so that income from sales is maximised from the smallest possible factory with the lowest possible investment in equipment. This is called "sweating the assets"; you make the equipment and factory run long hours to ensure that the quantity of product made is as high as possible.

The techniques for sweating the assets in manufacturing industry have been to work shifts so that the machines can be utilised for 24 hours a day for 7 days a week. The food industry is unique in that there will at some stage need to be a stop in production to meet the hygiene requirements of the product being manufactured. The hygiene stop could be every 4 hours in a high-risk area of a chilled food factory, or it could be every 20 days in a snack food operation, where the hygiene demands are less on a long shelf-life food product.

The method of maintaining customer focus is to set the shift patterns to meet the demands from the customer. It is not possible in this book to describe all demands on your business; patterns will have to be set based on local conditions, but here are some examples of the ways in which manufacturing shift patterns have been changed to meet the needs of customers.

Shift patterns

If orders arrive at 22:00 for product to be delivered the next day, it would be best to set the manufacturing operation to start at midnight and run through to 16:00 the next day with two 8-hour shifts followed by hygiene from 16:00 to 00:00. In this way manufacturing occurs as soon as the date code changes at midnight. This maximises the shelf-life of the product, which is very important to short-life/fresh food manufacturers. This pattern has the added advantage of allowing flexibility at the end of the shift, to delay the hygiene work and extend the working day using overtime to manufacture additional sales or recover from breakdowns or other delays during the day.

Orders for products will vary from day to day and week to week. For manufacturers who need to match production to daily orders, such as makers of short shelf-life foods, it is important that the factory is capable of coping with the size of the orders on a busy day, but that the factory is not overstaffed on days when the orders are lower. There are two methods of achieving this: either fill the factory up on the less busy days with products that have a longer shelf-life, or make sure that the staffing level exactly matches the size of the order.

Filling the factory up is useful, as it makes sure that all assets of the business are fully utilised.

Case study

A manufacturer of chilled pizzas, a short shelf-life product, was suffering from orders that would move up and down in several ways. The sales of the pizzas would vary depending on the day of the week, with orders at a peak on Wednesday and Thursday, but quite low on a Sunday and Monday. The orders were also greatly influenced by promotional activity. When the pizzas were on promotion, the sales could raise by 500% on the varieties promoted. The sales would also dip when competitor products were promoted. Add to this the rises and falls that occurred due to the time of year, and rises that occurred when there was a big football game on the TV, and you can begin to see the problem. The company had to have enough capacity to cope with the volume of the largest orders for its chilled pizzas, but did not want the factory to be idle during low sales periods.

The solution the factory came up with was to develop a brand of frozen pizzas that could be used to fill the gaps when the factory was not required for chilled pizza production. The frozen pizzas were long shelf-life and so could be put into stock in frozen warehouses. When the factory was busy with chilled pizza production, the customers of the frozen products could be satisfied from the stock held. The stock could then be built up again when sales of chilled pizzas were lower.

This mixing of long- and short-life products in the same factory is a good way of levelling out the production required and providing a steady demand. However, the creation of new brands in this way can take a long time and is not always successful. The other downside of this method is that it relies on stocks of frozen pizzas to make it work. During an extended period of low demand for chilled pizzas, the stocks of frozen pizzas would grow to very high levels. The consequences of being reliant on stocks of finished product to keep your business in balance will be described elsewhere in this book.

Flexible working agreements

The other method of matching production capacity to demand is to make the capacity of the factory variable by having flexible working arrangements with the staff, so that they are available when the factory requires them but do not get paid when the factory is short of work. That sounds a bit harsh, but there have been some very successful implementations of this type of working arrangement.

Questions

- List three ways in which your business tries to be flexible to the demands of your customers.
- How could this be any better?

Annualised hours deals

This is a working arrangement where people are contracted for a certain number of hours of work per year rather than the more traditional contract of hours per week. To soften the impact of a fluctuating pay packet, there are usually a minimum number of hours per week that will be worked and the flexibility is built into around a quarter of the hours. A worker on an annualised hours deal will be contracted for 1920 hours (that is the equivalent of 40 hours per week for 48 weeks, so 4 weeks holiday per year). During a period of low sales, the worker can be reduced in hours by 8 per week to a 32-hour week. In busy times, the worker can be asked to work up to 48 hours. During the year, the worker will complete all of the 1920 hours in the original contract. There will be rules within the scheme as to the notice that the worker will receive, of additional or less hours required.

To soften the blow of annualised hours variable working still further, some companies pay the workers on this type of contract a monthly salary of one-twelfth of the annual contract, no matter how many hours have been worked in the month. One downside of this kind of arrangement is in the administration of hours worked and the position of each employee against their annual contract total. Stories have been told of workers who have completed their annual total of hours two months before the end of the contract, because they have been working extra hours to cover absence.

Annualised hours deals need very careful setting up to ensure that there is full under-standing of the scheme and the implications of different business situations are fully thought through.

Uneven shift patterns

This method of matching production capacity to sales is based on setting up the shift patterns to match the peaks and troughs in the sales pattern. If you know that the produc-tion required on a Monday is always 30% less than on a Thursday, then the shift patterns are set to bring in 30% less staff and to run 30% less of the machines. Sounds simple but it is often difficult to exactly match the sales in this way. To run a factory there are certain jobs that are required, whether you are running one production line or five production lines. The other difficulty with this type of system is that it is based on sales forecasts. Forecasts can be inaccurate and as a result last minute changes to the shift pattern may be required to achieve a better match.

Casual labour

The use of casual or agency labour is widespread in the food industry and in particular unskilled jobs in short shelf-life factories. Casual labour is a source of extra peo-ple that can be switched on at very short notice and switched off again when not required.

The casual worker can be organised in several ways. The company can employ a bank of people on casual contracts that come into work only when told. There is a vision of these people sat by their phone at home waiting for a call from the factory but in reality they are probably very busy people, perhaps with their own business or engaged in childcare. They get a call and then make the necessary arrangements to get into the factory.

The second method of employing casual workers is through an agency or gangmaster. The workers are employed by the agency and are "rented out" to factories at an agreed rate to pay the worker, cover the costs of the agency and make a profit for the agency.

Casual labour could be the solution that allows a factory to exactly match its produc-tion capacity to the size of the orders. Casual labour does have addition risks however. Because of the casual nature of the worker, they will be very often untrained in anything but the minimum level of knowledge needed to work in a food factory. Some basic hygiene and safety training is often carried out along with some quick on-the-job training of what is required.

There will also be issues around motivation and integration into the production team. It is very difficult for a casual worker to perform at a high level in what are unfamiliar surroundings and working alongside unfamiliar people.

Continuity does not exist for most casual workers, so there is no gradual building of knowledge and skill and little, if any, opportunity to contribute to the team in terms of moti-vating and helping fellow workers or helping to improve the process in any way.

It has been known for some factories to attempt to run using levels of casual workers in excess of 50% during busy periods. This has a major impact on the performance of the operation, with no focus other than to "get the orders out on time and in full". This is sustainable for short periods of time but gradually working with high percentages of casual workers does start to erode performance improvement. A consequence of this lack of improvement momentum is that costs start to rise when those of your competitors may be falling.

In reality, the matching of production capacity to order volume is vital to food companies that cannot allow costs to be out of step with sales income. Most factories have developed plans and systems of operation that take lessons from all the examples above. Some include additional production for stock, some method of flexible working with core staff, and finally some degree of casualisation of the work force. A blend of these systems will improve the situation and allow the company to control its costs during both busy and slack times.

Contingency planning

Flexibility and responsiveness is not simply a matter of customer focus and flexible working arrangements. The final piece of the jigsaw is to ensure a supply of the correct quantity of raw materials to your factory.

The flexibility of suppliers to rises and falls in demand is an important part of a food manufacturing business. It is pointless having your bakery set up to respond to changes in demand if the flourmill cannot respond. All elements of manufacture have to have the correct level of responsiveness, without incurring additional costs. Raw materials, packaging, factory capacity, factory labour and distribution systems require the ability to respond quickly to a change in demand, so that the on time/in full requirements of the retailer can be met at the lowest possible cost.

In order to test the ability of your business to respond, it is normal to carry out a detailed simulation of possible scenarios and to take corrective action where weaknesses are found. This process is called contingency planning and allows for extensive "what if ... " questions to be asked of your business. What if there is a major increase in orders? What if our supplier has a factory fire? What if there is an outbreak of influenza among the factory staff.

Planning systems to enhance flexibility

The role of the planning function in the flexibility and responsiveness of a business is key. It is the people and systems that work out how the orders are to be met, that have the responsibility of delivering a workable plan to the shop-floor of the factory. The planners should have all information that they need to decide on the best course of action to optimise performance in the factory at the same time as meeting the requirements of the customer. There are many systems of planning that are used in the food manufacturing industry, from scribbled notes following a few calculations all the way to a fully automated computer-based system as part of a Material Requirements Planning (MRP) or

Enterprise Resource Planning (ERP) system. The information required to make a plan that works is always the same. Knowledge of all the factors and constraints of production is required. Product recipe and information about raw material stocks and deliveries. Manufacturing speeds and labour requirements. Customer orders and delivery requirements. All of this information is brought together into a plan that will work in the factory, if the factory delivers the required performance.

Flexibility and responsiveness starts with the planning function and requires the ability to think outside what is usual and routine. But the plan should not be unachievable. Setting a plan that requires the factory to achieve a higher than normal performance will lead to failure and additional costs, but it could also lead to disappointed customers.

Summary – Improving flexibility and responsiveness

This chapter looked at how some Lean Manufacturing techniques can help improve a business by impacting on flexibility and responsiveness.

Having systems that allow the business to respond quickly, without losing control over costs, is a major feature of food manufacturing factories. Customers demand flexibility and responsiveness, so if a food manufacturer wishes to stay in business and continue to make a profit, systems have to be developed.

The use of flexible working arrangements and casual staff in the food manufacturers is commonplace. Without these arrangements the on time/in full delivery requirements of the retailers would not be met or they would only be met by building in large quantities of slack into the systems. Slack is not a luxury most food manufacturers can afford, as they operate with very low profit margins.

Chapter 7

Improving what we do

The food industry is extremely competitive and in order to be successful there is a need to constantly improve the way the business is carried out. Staying still is not often an option when your competitors are able to take your sales by offering something better to your customers. This process of continuous improvement is one where the whole business is able and willing to look hard at what they are doing and try to find a better way of achieving the same thing. Better can be defined in many ways, especially in the food industry; it could be safer, it could be higher quality, it could be less wasteful, but most of the time better is less cost.

There are several techniques used to look at what is done in your factory and to try and point the way towards an improvement. Food factories tend to be pretty busy places and often a potential improvement can be missed because no one in the business can see the wood for the trees. Inefficient and wasteful methods continue for a long time, because no one has spotted the potential for improvement (see Figure 7.1).

Improving anything inevitably means that things have to change and this process of change can also be a block to improvements made. The important topic of management of change will be returned to later.

Continuous improvement – radical improvement by a thousand tiny steps

The constant striving to make things better is the driving force behind many successful food businesses. Improvements are rarely made in one big leap, although this can happen sometimes with a massive investment in new machinery and systems. Big improvements are more often made by the application of hundreds, if not thousands, of tiny changes carried out by several small teams in the business. But how can you tell where an improvement can be made? The answer to this one is simple; an improvement can be made to everything. Every task, every machine, every system, everything. The trick here is to try and pick the improvements that will give the best return for the effort that will be required to implement the change. "Where will we get the biggest bang for our buck?" or "Where will the company improve the most for the smallest input of time and resources?"

The final part of continuous improvement is to ensure that the changes that are made are well communicated and included in modified SOPs where required and are trained

Handbook of Lean Manufacturing in the Food Industry, First Edition. Michael Dudbridge.
© 2011 Blackwell Publishing Ltd. Published 2011 by Blackwell Publishing Ltd.

Figure 7.1 Where are the opportunities to improve in your factory? Sometimes the improvement opportunities are clear, but most of the time it is difficult to see the root cause of an issue.

out. In this way the improvement is not cemented into the business until the staff are retrained and the improvement is embedded into the way that task is carried out.

A great place to start looking is the DWOR. This will highlight areas where work is required and will prompt investigation into the reasons behind the numbers. The DWOR is useful in two ways to show opportunities. It is made up of the KPIs for the business, so is right on the button of the important things in the factory. The DWOR also contains targets, and failure to meet targets (or indeed to always exceed targets) is a sure sign that things are not what was expected.

Case study

The DWOR for a department in a factory reveals that the level of waste is varying between 0.8% on a good day and 2.9% on a bad day. The best result ever was 0.2%, which was achieved 3 weeks ago.

Investigate what is causing the high levels of waste on the days of poor performance and there will probably be four or five improvements that can be made.

Could be poor stock rotation, could be poor quality ingredients, could be machine breakdown, could be ...?

The likely outcome is that the high waste was caused by more than one factor and it is therefore logical to go for the easy ones first. "Pick the low hanging fruit." "Go for the quick win." Once the easy issues have been solved and improvements made, revisit the issue armed with new data and pick the more difficult areas to work on.

The DWOR will reveal performance history for all KPIs. Armed with that information, investigations will reveal the reasons behind poor labour efficiency, poor quality scores,

Figure 7.2 The process in your factory may be sub-optimal and in need of some attention, to ensure that it is carried out at the minimum possible cost.

high absenteeism and high levels of breakdowns. It is only after the investigations into the causes that the team can come up with, and try, some improvement ideas.

Process development

This is a technique where a production process is analysed and opportunities to improve the process in some way are identified. This is an area that is always full of opportunities for improvement. If the process is relatively new, it will have been put together during the factory trials of the product as it was prepared for launch. The process, by definition, will have been developed on relatively short production runs, using materials bought in small quantities and maybe with mock-up packaging. Maybe some of the machinery used in the process was relatively new and not well understood during this launch phase.

Once the product is launched and longer production runs are scheduled, larger batches of raw material are sourced and processed in larger batches, and correct packaging materials are being used. More data on the process is available than would have been present during those early trials. The product development team have moved on to their next new product and so it is possible that the process used in the factory is not optimal and there will be many improvements that could be made (see Figure 7.2).

Figure 7.3   Storage is a non-value adding activity. Cost is added to the product but no value from a consumer point of view.

If the process has been in place for some time, it is hoped that it has been developed to a point where it is close to optimum for the product being manufactured. However, this is not always the case. Long established processes may well have been "tinkered" with over the years, in response to longer-term issues such as raw material variation or mechanical issues on the machinery used in the process. As a result of this tinkering, the process may have developed some instability, which causes it to "fall over" on occasions and produce reject product. A relatively minor issue on the production line causes a major change in the product being manufactured.

It is a bit like being near the edge of a cliff; a process that has fallen over the edge and is making rejects is put back onto the edge by minor tinkering. The real cause of falling over the edge is never discovered. The next time the process falls over the edge, more tinkering is carried out and again the process is put back on the edge with no real investigation. This continues for some time, with the process becoming less and less stable and unable to cope with even minor issues without creating reject product. It takes thorough investigating to get to the root cause of a problem and tinkering with a process will ultimately lead to poor performance with compensating errors built in.

Some companies have recognised this area and employ a Process Development Technologist to work with the production teams to ensure that the products are performing as originally planned. Meanwhile the production teams are trying to manage a process that is producing poor yields, high waste, erratic quality or slow run speeds. Once again these factors can be identified in the DWOR, but other sources of information could be materials and packaging over usage, short delivery to customers or even the fact that the people on the shop-floor do not like running the product because "it always gives us loads of hassle".

Suppose our process is out of control and reject products are produced. The biscuits are too big for the packaging, the pizzas are oval, the yoghurt is too runny, and the sausage rolls are too short. It is a bit like a car falling over a cliff. What is needed is to get the car back onto the top of the cliff and then deciding what should be done to move it back from the edge so its will not fall off so easily next time?

How the process can be developed is outside the scope of this book. However, there are some rules of process development that can be applied simply to improve matters:

1 Make sure the whole process is considered when making changes. Developing a process is a bit like trying to squash a fluffy pillow into a bag that is just a little too small. You squeeze in one area and the pillow pops out elsewhere. In process development, you improve the yield and the quality suffers, you make the process faster and the yield disappears, you reduce the headcount on the production assembly line and the giveaway goes up, you reduce the waste at the start of the process and the waste at the end of the process goes up! It is important that all parameters of the process are monitored to ensure that improvements in one area do not result in a worsening situation elsewhere. Sometimes though, it is possible to accept a worsening of performance elsewhere, if the overall process is more efficient.

2 Ensure that the process remains under control by checking the calibration of instruments used to measure the process. Scales, temperature gauges, checkweighers, humidity and pressure gauges, mixer timers and conveyor speed controllers all need to give accurate information for the process to be controlled and developed.

3 Just change one thing at a time if possible. Make sure that the process development is logical and that time is given for the results of changes to emerge in the KPI data. It is then possible to say that a change in the process resulted in an improvement. If more than one thing is changing about the process, it is then difficult to make the link. Changing one thing at a time can be slow, especially if the results of the change cannot be measured for some time, say on a long cooking process or on a blast freezing process. Where there is a risk that the product may not be sellable as a result of the process changes made, you should proceed with caution. This, again, is important on longer processes where a lot of product could be wasted.

4 Communicate. It is important that everyone knows that you are going to move away from the SOP to try to improve the process. Involve others in the decisions to ensure that their knowledge of the process is included in your thinking.

5 When a process change does result in an improvement in yield, a reduction in waste or an improvement in quality ensure that the change is correctly communicated. Rewrite the SOP, change the factory paperwork, retrain the staff, do everything to ensure that the improvement sticks.

6 Audit the processes in the factory on a regular basis, to ensure that they are being run to standard. Compare what is happening on the shop-floor with what should be happening. In this way the discipline of running to standard will be enforced, but also it could prompt further process improvements when the process is not run according to the SOP. A simple question of "Why?" addressed to the process operator may be enough to identify a process issue.

7 Keep records of the process, so that a history of process settings is built up along with reasons why the process was altered. In this way your business will build up a huge bank of knowledge about your process and this will be able to guide your thinking in the event of the car falling off the cliff again.

Questions

- How are processes developed in your factory?
- Who is responsible for coming up with improvements in the way things are done?
- Are there examples of long-running issues in your process that have not been tackled and generate problems on a regular basis? Write down a couple while they are fresh in your mind.

Value stream mapping (VSM)

This is a great way of identifying potential improvements in the efficiency of a food factory. The technique relies on the idea that raw materials arrive at a factory and are relatively low-value items. As the materials are put through the factory, they gradually gain in value to a point where the product sent out to the customer is higher in value that the raw materials used to make it.

The trick in manufacturing is to add value, from the viewpoint of the consumer, in a way that costs less than the value that is added. That way the business will make a profit.

VSM looks in detail inside the factory, at each step in the manufacturing process and identifies where that value is being added from the consumer viewpoint and where all of the costs are being incurred.

Case study

Let's have a look at a factory making pizza

The factory buys in the raw materials. Frozen pizza bases, blocks of cheese and tomato sauce (this is a simple pizza!). The task of the factory is to bring these components together to make a pizza of higher value than the costs incurred.

Let's look at the pizza bases; they arrive at the factory in cardboard boxes of 50 bases, individually wrapped. They are purchased by the lorry load of 26 pallets, that is 52,000 bases and are off-loaded using a forklift truck and placed into the cold store. 52,000 bases will last the factory one week. They are taken to the head of the production line, are deboxed and unwrapped and placed onto the conveyor ready for the sauce to be added.

Let's have another look at that process from the consumer point of view. Does the consumer know that the bases are bought in ready baked? And if they did know, would they think that added any value to the end product. If they were imported from Italy, then maybe that would add value, otherwise no added value here, only cost. The VSM for the flow of pizza bases to the head of the line shows no added value. The task is a simple one then; we must minimise the cost of providing the bases to the line and anything goes in this case, as no value is added from the consumer point of view.

Some improvements could be achieved by having the bases delivered in bulk rather than individually wrapped. This will reduce the cost of the bases, assuming the supplier costs are lower because they no longer have to wrap them. But taking the wrapping away will have an effect on this frozen baked product; freezing and storage will dry the product out, which is something the consumer will notice and will consider it to be a reduction in value. Removing the individual wrapping is not possible unless the pizza bases can be protected in a different way. How about wrapping them in 4s, 8s or 50s? The protection would be there but the costs would be lower to deliver exactly the same base to the assembly line.

Before we move on to the next component, let's just summarise the cost savings that could occur with the simple packaging change. Costs would be reduced from the supplier, because their cost would be lower. Costs in your factory would be reduced, because the opening of the individual pizza bases would not be required, so there is less manual input into the process. If the bases are wrapped in 50s, for example, only 1 bag has to be opened rather than 50 bags, which is quicker, easier and cheaper.

There is much more that could be done. We employ a forklift driver and have a cold store space currently used for handling bases. Is there anything about the delivery and storage method that could be reduced or eliminated?

If we were able to take a daily delivery of bases, say five pallets per day rather than a whole lorry load, the forklift driver could deliver them straight into production rather than have to put them away into storage. This would allow the cold store to be switched off, or the space reallocated, and the workload of the forklift driver would reduce. Once a pallet of bases is picked up on the truck it is delivered to the point of use rather than be placed into storage, only to be picked up again later (see Figure 7.3).

Handling and storage

The elimination of double handling is a key saving that is often identified in VSM exercises. Handling and moving items around a factory adds no value from the consumer point of view. Storage of items is also not a value adding activity. All that transport, handling and storage achieves for an item is to add cost and no value is *ever* added.

There is more about the handling and storage of items that adds unnecessary cost. When items are moved, as well as the cost of the movement, truck rental, driver and energy costs, there is the additional cost of administration and losses from that process. If you rent a truck, someone has to pay that rent, check the invoice and manage that supplier; and how about the safety costs, high visibility jackets, safety inspections, etc. Other administration includes labelling and traceability of the product being moved. If an item is moved inside a food factory, it will usually have to be issued from one area to another, the date and time of movement will have to be recorded, and then someone will have to inspect those records and file them away for future reference. If the handling movement is eliminated inside a food factory, it reduces costs in all sorts of ways.

The same things apply when an item is stored. A space to store must be created and there is a cost associated with that. Assuming the space is already available, lighting, and heating or cooling have to be provided. The storage areas have to be cleaned and

Figure 7.4 Storage areas are not cheap places. A lot of work is done to minimise the cost but the ideal solution from a Lean Manufacturing point of view is to eliminate the storage altogether. No storage means no storage costs and no double-handling costs for that material.

pest control measures applied. They have to be maintained and, perhaps most impor-tantly, the stock has to be counted daily or weekly and the stock rotation managed in such a way as to ensure that materials are not wasted. From a VSM point of view, the value of the item coming out of the store is exactly the same as its value when it went in, so storage adds cost but does not add value (see Figure 7.4).

At this stage you are probably thinking, "I cannot move anything, I cannot store any-thing, how can my factory run?" As with many of these techniques, perfection is not possible or financially viable. Some double handling will be necessary, along with some storage. The aim of these exercises is to reduce these non-value-adding activities to an absolute minimum and if you see these activities occur it should ring alarm bells and get you thinking of ways they can be reduced.

Case study

We have looked at the pizza bases and learned a few things about VSM. Now look at the supply of cheese to the pizza assembly line and see if there are more principles we can learn.

The cheese for our pizzas is supplied in 20 kg blocks. The blocks are wrapped in plastic shrink wrap and stacked onto a pallet. On delivery they are off loaded and placed into the chill store, while microbiology checks are made; this takes three days for the results to come back. Once the cheese has been given the all-clear by the technical department, it is released to production and is issued to the department who pass it through a large grating machine. It takes 30 seconds to grate a 20 kg block and the grated cheese is placed into plastic tubs. The tubs are labelled with batch codes and dates of grating and placed into a chill store ready to be used on the line. When the line requires the grated cheese, it is issued to them and the batch codes are recorded to ensure traceability of the cheese onto the assembly line.

The tubs are emptied into the machine that weighs the cheese onto the pizzas as they flow down the assembly conveyor. The machine uses 20 kg in 3 minutes.

As already learned, double handling, transport and storage are not value added activities and it can already be seen that these occur in the cheese process. There is also administration of the cheese in terms of traceability that does not add value.

What is needed is a closer look at the timings in the cheese grating process, to see if value is added?

First, there is the three days' delay waiting for microbiology clearance. Is this adding value? The process of checking if all is well is not a value added activity, as no quality checks add value; they just check that the job was done correctly in the first place. The microbiology check is essential from a consumer safety point of view, but does not add value. Is there any way that this check can be done without incurring any cost? Supplier Assurance is a system by which your supplier is not allowed to deliver materials to your factory that have not been checked fully to meet your specification. Why should you pay for a check to see if your supplier has done the job correctly? And why should you accept a delivery of materials that has not been tested to check that it meets your specification. Supplier Assurance is a system in which the material arrives with a Certificate of Conformity (a C of C). This certificate states that the material meets your specification, so no further testing should be required before the material is used.

Remember, this system would stop us having to store cheese for three days while waiting for the results of the checks, so there is a cost reduction. It would also allow cheese to be delivered and issued direct to production; the same savings seen with the pizza bases earlier are starting to be identified.

The cheese is on site and is now going to be grated for use on the pizzas. It can be grated at the rate of 2400 kg per hour. It can be used at the rate of 400 kg per hour and grated at 6 times the rate of usage. The process of grating cheese from a VSM point of view does add value; the pizza would be seen as inferior is if arrived with one lump of cheese on the top. What opportunities exist to reduce the cost of adding that value? There will be an opportunity to more closely match the speeds of the grating and depositing processes. Where two processes are linked, but are carried out at greatly differing speeds, there will be a tendency to build buffer stocks and for the efficiency of the supplying operation to reduce as the need and motivation to maintain a pace will not be there (see Figure 7.5).

The ideal situation is that the output of the cheese grating operation exactly matches the demand from the cheese weighing system. A small quantity of buffer should exist between

Figure 7.5   The ideal solution would be to match the supply and demand on the production line (picture courtesy of Ishida Europe).

the two processes to allow for any short stoppages on the grating machine but otherwise the cheese should be grated Just In Time (JIT) for its use on the weighing system.

Just in time deliveries and processing

JIT is a disciplined approach to manufacturing that allows for materials to arrive at their point of use just as the previous batch is coming to an end. This minimises the buffer stock between processes and as a result the control and storage of that buffer stock is also minimised. JIT systems are also a way of controlling incoming raw materials from your supplier, to ensure that stock held at your factory is at as low a level as is viable.

JIT systems work well when there is a high level of reliability in the supply systems. If supply systems are in anyway unreliable, JIT becomes Just Too Late (JTL)! The consequence of a poorly performing JIT system is that production lines grind to a halt because of a lack of the raw materials required. Production efficiencies drop and the costs of production rise rapidly. However, the costs of running a production system, which is full of buffer stocks and raw material control issues, can be equally damaging to the profit margin; the costs are just more deeply hidden.

What are the possible solutions to the supply of grated cheese to the pizza production line? The ultimate solution would probably be to mount the grating machine immediately above the weighing machine. As the weighing machine requires more materials, the grating machine is switched on and one block of cheese is grated. The grating machine is then switched off to wait for the next call for cheese. In this way cheese is only grated when required.

Analysing this possible solution makes sure that problems are not caused for the business by trying to reduce stock and move towards JIT grating of cheese. The way to analyse this type of improvement is to first list the advantages and possible cost savings, but then to consider all possible disadvantages of the change.

An advantage is that it is part of the positive thinking method that needs to be in place in a factory undertaking continuous improvement. If you consider the downside first, you will have talked yourself out of the proposed change, even before you have considered the upside.

Advantages of JIT cheese for our pizza line

- Reduced double handling and manual handling leading to headcount reduction
- No need for storage space for grated cheese
- Reduced stock rotation and stock counting issues
- No need for tubs to contain the grated cheese, so less to clean
- Safer – no movement of grated cheese from store to the line
- Reduced waste
- Easier traceability.

Disadvantages of JIT cheese for our pizza line

- Capital investment required to create the new configuration
- Risk of cheese grater breakdown stopping the line very quickly
- Reduced flexibility
- More difficult access to grater for blade changes and maintenance
- Increased chance of foreign body contamination.

Questions

- Write down an example of where JIT is used in your business.
- Now write where JIT would show a benefit, but is not used.

Once the pros and cons have been itemised, the next step is to try and place a monetary value on them so that a judgement can be made. The way in which a monetary value is decided is different from company to company, so this book cannot give much guidance. However, some rules will apply that will be the same in most food companies.

Labour costs are usually calculated as the whole cost of employing a person rather than just the wages costs. Costs such as holiday pay, sick pay, employer national insurance and health insurance costs are included along with items that are easily overlooked, such as costs for protective equipment and overalls, subsidised canteen and even the cost of supervision of that person. The true cost of someone in a factory can often be as much as twice the hourly pay rate.

Risk Analysis

Risk Description	Likelihood	Consequence	Risk Factor
	Score 1 to 10	Score 1 to 10	Likelihood x Consequence
Foreign body contamination (before)	1	6	6
Foreign body contamination (after)	2	6	12

Figure 7.6   A method of risk analysis is required to check that the proposed change has been well thought through.

Analysis of improvement project risks

A project such as this will either reduce or increase the chances of an event happening; for example, in this case there is an increased chance of foreign body contamination and a reduced chance of stock rotation issues. These situations are difficult to quantify and place a value against but some systems have been developed. Risk analysis is all about the chances of an event happening, the likelihood and the consequences if the event does occur, and the impact. It is normal to allocate these two factors a number from one to ten. The numbers are then multiplied together to give a risk factor. Let's take our foreign body contamination issue. The likelihood of the issue is slightly increased from say 1 to 2 when the grater is in its new position. The actual contamination is the same, but the new system has less opportunity for the contamination to be detected. The consequence is the same and quite high at maybe a 6. The risk factor for this part of the analysis has moved from a 6 (1 × 6) to a 12 (2 × 6). The financial value of that risk has doubled. Analysis of the cost of contamination from the grating machine over the last year will give a value of the current cost per year; the new system could be expected to double that cost (see Figure 7.6).

Once all costs and benefits of a project such as this have been analysed, it would normally be expected in the food industry for the capital investment to be repaid by lower costs within one year. Again the rules on payback requirements will vary from company to company. It is normal for companies manufacturing chilled, short-life foods to demand very short payback periods. Companies in more stable, longer-life foods may be able to work with projects that take longer to pay back the capital investment through reduced costs.

Ultimately, a decision has to be taken whether to make the change or not. In the case of bigger projects involving capital, this can take some time because of the need for the finance to be approved and the equipment to be purchased and delivered. It is often a lot quicker to make an improvement that is no cost or low cost. For example, painting a few lines onto the floor of a work area to help with workplace organisation, marking a gauge with a red zone to indicate when the process is out of control, or organising a desk so that it is neat and tidy and every shift can find the paperwork required with the minimum of fuss. All these items will have an impact on performance but will, at most, cost pence to carry out.

This VSM technique will help us identify areas where cost is added to the product, but the value of the product does not increase, or the cost increase is greater than the value increase. The last component of our pizza study will show if there are any further points to learn.

Case study

The sauce is bought in ready to use on the pizzas. This is a very simple pizza we are making here. The principles we have learned with the bases and the cheese can still apply to this simple situation. The sauce is bought in large quantities and stored on site; what are the opportunities there?

The sauce is issued to production (admin and tracability costs will usually be incurred here) and then it is put into the depositing machine from the 15 kg tubs it was delivered in. The weight control of the sauce is important to the business, as the sauce is the most expensive component of our pizza. We already know that checking production is not an activity that adds value, so how can we check that the correct quantity of sauce is used without incurring additional costs? Use inline automatic checkweigher with automatic feedback to the depositor, so that if the weight at that stage of the process is incorrect, the depositor will be automatically adjusted (see Figure 7.7).

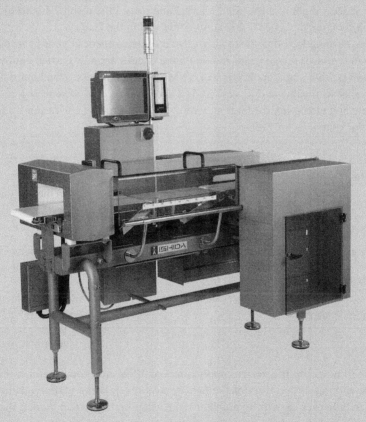

Figure 7.7   An inline automatic checkweigher reduces the cost of the non-value adding activity of checking the weight of a product. it can be linked to a depositor, which automatically adjusts the weight to ensure it is always correct (picture courtesy of Ishida Europe).

The final outcome of VSM, which gives rise to a lot of improvements in performance, is where a cost cannot be avoided but it adds no value to the product. This would be a Critical Control Point (CCP), where a check needs to be made to ensure the product is safe and legal to sell. These CCPs occur all over food factories and are essential to the management of the process. Where a check of this nature is essential, it must be done. The only way to do the check and avoid adding cost, is to combine this non-value adding activity with an activity that does add value.

For example, if our sauce depositing weight was a CCP to ensure that the pizza met the specification and was legal to sell (maybe our pizza is a healthy option and makes nutritional claims), the check on sauce weight could be carried out by the person who was adjusting the sauce spreading machine. That role would add value as it is improving the value of the product from a consumer point of view. Part of that role would now be to check the sauce weight at a specified interval and record the result. This would be a no cost option for covering a non-value-adding role within the factory.

You may have noticed that the quality assurance department in your factory has been shrinking in size over recent years and that more checks are now carried out by production operatives; this is the application of VSM in your business. Combining non-value adding checks with value adding activity is a common method of reducing costs, while maintaining and increasing the level of control over the product and responsibility of the operative for what they are doing.

Questions

Think where else the powerful technique of VSM could be applied: eliminating, reducing or combining non-value adding tasks so that value adding and cost adding tasks become more closely aligned:

• Do any areas of your factory spring to mind, where cost is added but the value does not change?
  ○ Think about storage.
  ○ People moving items to a new location only for the item to be moved again later?
  ○ Double handling?

There will never be the perfect application of VSM in a food factory; the cost of doing that and the inflexibility of the factory as a result mean that no business could justify that. However, as with so much of Lean Manufacturing techniques, the journey towards perfection is what brings the continuous improvement that food businesses desire.

Reducing the processes that do not add value

The business of food manufacture is a very competitive one. The need to minimise cost is a technique that allows food factories to continue in business and supply the consumer with what is required.

The concept of value is one that is constantly changing. How much is something worth, its value, is very much in the eye of the consumer. A few years ago high-quality packaging of food was seen to enhance value and allow the consumer to justify paying a higher price. In recent years, over-packaging has been seen as something that can detract from the value of food products.

Washed potatoes added value but now some of the highest priced potatoes in the retailers are unwashed organic potatoes with some skin blemishes.

In a world where the value of everything is changing, it is important to stay close to the consumer and make sure that your products reflect the value that is placed on them.

There is a great opportunity in food retailing to grow a business very rapidly if your products closely match the value given to them by the consumer. The value of a pre-prepared and cooked ready meal for microwave reheating is reducing as consumers move towards freshly cooked ingredients in their homes.

Once you have designed a product to meet the value expectations of the consumer, there is still the task in the factory of manufacturing that product and carrying out the VSM exercise to reduce or eliminate the processes that do not add value.

The final step in the process of improving what is done is the one that will improve labour productivity in a production process.

Line balancing to optimise labour productivity

When a team of people are working together to manufacture a food product, it is important that the workload is shared as evenly as possible among the team members to ensure that productivity is optimised. To get the maximum benefit from working as a team, the task of manufacturing the product is often based around an assembly line.

The food is conveyed at a steady rate down the conveyor belt. As the food passes a workstation, the operative on that station carries out a task. The food then passes the next station and the next operation takes place. The food could be a multi-component ready meal or a layered salad product, a tin of assorted biscuits or a fancy cream cake (see Figure 7.8).

For example, the assembly of a high-quality chilled pizza; here are the work stations along with the times that each task would take a skilled operative to carry out (see Figure 7.9).

You can see from the above table that the slowest task on the line is the counting and placement of the pepperoni slices at 7.5 seconds. If there was one operative per workstation, the maximum number of pizzas produced would be 8 per minute, because the pepperoni person would hold up the rest of the team. The labour productivity of the team of 9 people would be 0.88 pizzas per person per minute.

How can we change things to improve the productivity of this operation?

There are a couple of choices:

1 A second person could be put onto pepperoni, which would reduce the time required to count and place the pepperoni to 3.75 seconds, as 2 people would be doing it.

Figure 7.8 The balance of labour on a production line will allow the output to be as high as possible, with each member of the team contributing as equally as possible.

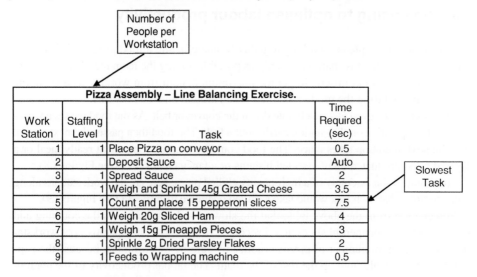

Number of People per Workstation

Pizza Assembly – Line Balancing Exercise.			
Work Station	Staffing Level	Task	Time Required (sec)
1	1	Place Pizza on conveyor	0.5
2		Deposit Sauce	Auto
3	1	Spread Sauce	2
4	1	Weigh and Sprinkle 45g Grated Cheese	3.5
5	1	Count and place 15 pepperoni slices	7.5
6	1	Weigh 20g Sliced Ham	4
7	1	Weigh 15g Pineapple Pieces	3
8	1	Spinkle 2g Dried Parsley Flakes	2
9	1	Feeds to Wrapping machine	0.5

Slowest Task

Figure 7.9 A basic data capture table to record the times required for each task when carried out by a skilled operative.

2 Or some of the other tasks could be combined to allow people to stay busy while they are waiting for the pepperoni person (see Figure 7.10).

Option 1 would be to put an extra person on pepperoni. The increase in staff numbers by one person means that now the slowest operation is the weighing on of the sliced ham, which takes 4 seconds. In this system the maximum output has increased to 15 pizzas per

Pizza Assembly – Line Balancing Exercise.			
Work Station	Staffing Level	Task	Time Required (sec)
1	1	Place Pizza on conveyor	0.5
2		Deposit Sauce	Auto
3	1	Spread Sauce	2
4	1	Weigh and Sprinkle 45g Grated Cheese	3.5
5	2	Count and place 15 pepperoni slices	3.75
6	1	Weigh 20g Sliced Ham	4
7	1	Weigh 15g Pineapple Pieces	3
8	1	Spinkle 2g Dried Parsley Flakes	2
9	1	Feeds to Wrapping machine	0.5

One extra person reduces the time taken on pepperoni

Figure 7.10   An attempt to make the line more efficient with the addition of one extra person.

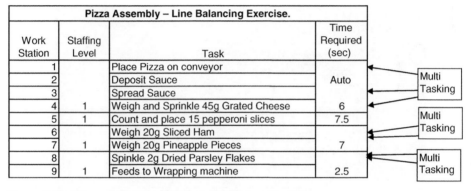

Figure 7.11   The combination of tasks to improve productivity.

minute. The labour productivity has moved up to 1.5 pizzas per person per minute. That one move has very nearly doubled the productivity.

Option 2 is to combine other tasks to reduce the headcount on the line and balance the line using that method (see Figure 7.11).

The crew on this line has now been reduced to four from the original eight. The line is better balanced, with more even allocation of tasks. The slowest operation is still the counting and placing of pepperoni at 7.5 seconds, so the output is 8 per minute. With 4 people, this gives a productivity of 2 pizzas per person per minute. It can also be seen here that the operative on workstation 8/9 is still under-occupied for 5 out of every 7.5 seconds.

The optimal balance?

The optimal line balance will be a line where everyone is fully occupied for the entire time; in this case that would be as shown in Figure 7.12.

In this line balancing exercise, all tasks are reduced in duration to the time required for the quickest task, 0.5 sec; by the addition of people to each of the workstations. Pepperoni counting and placement takes 1 person 7.5 seconds or 15 people 0.5 seconds. The staffing on the

Pizza Assembly – Line Balancing Exercise.			
Work Station	Staffing Level	Task	Time Required (sec)
1	1	Place Pizza on conveyor	0.5
2		Deposit Sauce	Auto
3	4	Spread Sauce	0.5
4	7	Weigh and Sprinkle 45g Grated Cheese	0.5
5	15	Count and place 15 pepperoni slices	0.5
6	8	Weigh 20g Sliced Ham	0.5
7	6	Weigh 15g Pineapple Pieces	0.5
8	4	Spinkle 2g Dried Parsley Flakes	0.5
9	1	Feeds to Wrapping machine	0.5

Figure 7.12   Reducing the time for all tasks to the time for the quickest task.

Figure 7.13   I just wanted an excuse to show you a picture of a lovely pizza! It is an example of a multi-component product that is often made by large teams of people at a very high production rate.

line has increased to 46 people and the output has increased to 120 per minute. The optimal productivity is 2.6 pizzas per person per minute. This is achieved with all staff being fully occupied. The very high number of staff required for this balance is likely to be an issue. To fit all of these people onto one conveyor system may well need more space and a bigger factory! Also an issue would be, at this run rate of 120 pizzas per minute, can the sauce depositor keep up, can the wrapping machine keep up, and is the canteen large enough?

Finally, at 120 pizzas per minute, does the factory have sufficient work to keep these people occupied for a full shift (see Figure 7.13)?

The line balance that is most likely to be adopted is one that is not optimal but accepts the fact that there is some idle time on some of the workstations. The final picture would probably look something like that shown in Figure 7.14.

Pizza Assembly – Line Balancing Exercise.			
Work Station	Staffing Level	Task	Time Required (sec)
1		Place Pizza on conveyor	
2		Deposit Sauce	Auto
3	1	Spread Sauce	2.5
4	2	Weigh and Sprinkle 45g Grated Cheese	1.75
5	3	Count and place 15 pepperoni slices	2.5
6	2	Weigh 20g Sliced Ham	2
7	2	Weigh 15g Pineapple Pieces	1.5
8		Spinkle 2g Dried Parsley Flakes	
9	4	Feeds to Wrapping machine	2.5

Figure 7.14 A balanced line with a reasonable allocation of tasks and numbers of operatives.

This balance has 11 staff producing pizzas at the rate of 24 per minute and gives a productivity level of 2.18 pizzas per person per minute. The slackest time is on the pineapple task, so it would be worth looking at the methods used on this task to see if it is possible to improve them to hit the target time of 2.5 seconds. If this could be achieved, then the staffing on pineapple could be reduced to 1 and productivity for the line would move to 2.4 pizzas per person per minute.

You can see from this example that line balancing is a tool that can be used to model different line set-ups and calculate the most effective use of staff. It is just a tool though and needs to guide you to higher productivity in line assembly situations rather than be the answer to all your productivity needs.

Questions

• Have you got an application for this type of technique?
• It is very powerful and forces you to look at what the team are doing, even before you take the measurements of the time it takes them to do it. Try an exercise on one line and see if it shows any advantage.

Summary – Improving what we do

This chapter has looked at the ways in which some Lean Manufacturing techniques can be applied to making improvements in what is done.

By carrying out some process development in your factory, you could reduce the costs by eliminating process steps or at least looking at each step from the viewpoint of how much it costs to carry out. The concept of seven rules for process development has to be understood before we can look at the powerful technique of VSM.

VSM is a method of looking at what you do in your factory from the viewpoint of your customer. If a part of the process adds no value from a customer point of view, and simply adds cost, then that part of the process should be changed to reduce or eliminate that step.

Simple steps such as storage of an item adds cost but does not add value and the conclusion is that all storage has a non-value adding activity and should be eliminated whenever possible.

Finally, this chapter introduced a technique called Line Balancing, which is used in manual assembly operations to ensure that the work is efficiently allocated and the people are equally occupied.

Chapter 8

Improving how we do things

The previous chapter focused on improving the way things are done. That is a great way of making improvements in a food business, but it is not the only way. An area that needs to be considered is *how* things are done, rather than *what* is done.

Try to think of an example on *how* things are done in your business. For example, the production line did not start on time this morning, because there was water in the electrical panel from the hygiene operation.

Questions

- How would your business go about sorting out that issue, so that it never happened again?
- How would the issue be raised?
- Who would make the decisions?
- Who would actually do something, and what would they do?
- What would be done in this sort of case?

I guess that your answer will have included a lot of different suggestions, but is your business consistent about how it handles situations such as this?

Here are some of the areas where thinking about the *how* rather than *what* will lead to a better food business.

Can the same things be done more quickly? Can the time it takes to launch a new product be reduced? Can the effectiveness of hand-washing be improved without increasing the time it takes? Can the plan for tomorrow be definitely available by 16:00 each day? Can the time it takes to change a production line from one product to another be reduced?

There are three ways in which a food business can improve how it tackles issues and ensures that things are improved efficiently and in a way that ensures that the issue does not reoccur.

Handbook of Lean Manufacturing in the Food Industry, First Edition. Michael Dudbridge.
© 2011 Blackwell Publishing Ltd. Published 2011 by Blackwell Publishing Ltd.

Making changes stick for all shifts

One characteristic of the food industry, that makes it different from other manufacturing industries, is that the manufacturing occurs around the clock and often seven days a week. This means that a large number of people are working in the same factory and on the same production process, but these people do not meet. The 24/7 basis of production also means that an issue that occurs for one team may not be discovered until after they have finished their shift. Working in shift patterns has advantages for the output of the factory and the profitability of the business, but it has disadvantages when it comes to the management of people and communication.

One area of concern that is often seen is an inconsistency in the methods used by each shift in the operation of the process. This was considered earlier, when looking at the use of SOPs. What about inconsistency in *how* things are done in different areas of the business?

In your factory, do all shifts allocate holidays in the same way? How about absenteeism? Is that controlled consistently by all the shifts?

Questions

- What are the implications for the business of these two areas being carried out differently on different shifts or in different departments?
- Have a think about your factory. Think of the last time you noticed that things were being handled differently in different areas of the business.

It is very difficult to be totally consistent across all parts of the business and total consistency may not be necessarily an advantage anyway. But it is a good thing that performance and methods are the same and that includes the *how* as well as the *what.*

Consistency of management controls is vital if the business is to optimise its performance. Staff in a business will tolerate strict rules and procedures required in a food business, but only if they are applied consistently for all.

For example, an error is made by a skilled operator on Red shift and as a result a large quantity of food is wasted. An investigation shows that the standard procedure was not followed and as a result the operator was given a disciplinary hearing and was issued with a written warning. The operator appeals against the warning based on the fact that the same error was made on Blue shift and no one was warned.

The management time taken in investigating and holding disciplinary interviews would have been better spent developing improvements to the process and motivating staff rather than dealing with the consequence of inconsistencies across the shifts. The need for the *how* as well as the *what* to be the same across all parts of the business is obvious, once you consider what happens to teams of people who are managed in different ways.

Consistency is vital in the day-to-day management of a food business.

Clear rules

To ensure that consistency exists, a clear set of rules and procedures is an obvious first point. This means that the rules and procedures need to be written down and understood by everyone. It is also vital that changes in the rules and procedures are recorded and new versions are issued. Document control in a food factory is one of the cornerstones of the management system. It is vital that if a change is made to a product, a process or a procedure, that new documents are issued and the old documents are removed and destroyed. Document control is carried out to support the work of the factory in ensuring that all systems and procedures are up to date at all times. Document control is often seen as an inconvenience in a factory environment, but without it the factory will find it very difficult to perform well and errors would occur.

Questions

- What are the systems of Document control in your factory?
- Do they work effectively?
- When was the last time you encountered a problem in your area that was caused by an out-of-date document being used?

Document control

Document control is a major part of a Factory Audit when your customers come and look at your factory. Controlled documents are sometimes limited to those issued by the Quality Department. How far does your Document control extend?

The following will be under a system of control:

Recipe control sheets

These list the weights of ingredients needed to manufacture a product. The control document may also carry information on the process conditions required. They are sometimes used to verify that the correct ingredients were used in the product and that the batch codes are captured for traceability records. The days of an operative having a little black book with all the recipes in have long gone in most food businesses. The little black book was never a controlled document and mistakes and recipe changes would result.

Process control records

Process control records are a way in which the required settings for a process can be communicated to the shop-floor, and again a way of the shop-floor recording the actual conditions in the process. It is vital that if a process change is required, maybe as the result of some work by an improvement team, that this document is updated and all old versions are removed from the factory.

The list of controlled documents that have implications for the quality and safety of the products manufactured can be a long one:

- Hygiene signoff sheets
- Pre production check sheets

- Slicer blade check sheets
- Pack weight check sheets
- Date code check sheets.

There is a cost involved in the flow of this information, which some food businesses are starting to tackle. The use of computers to help with communication of this information is increasing. Computerised data capture systems on the shop-floor have allowed for this communication to be automated or semi-automated. Supervisory, Control and Data Acquisition (SCADA) systems can help in the reduction of paperwork and the improvement of information flow in a business.

Questions

- What controlled documents do you have in your area or on your line?
- How do you know if the document you are looking at is correct?

How about these? Are the documents controlled to ensure that only up-to-date information is available on the shop-floor?

- Today's production plan
- Stock check sheets
- Waste recording sheets
- Standard run rates and staffing requirements
- Engineering maintenance schedules
- Conveyor belt widths and lengths
- Standard procedure to deal with liquid spillages
- Fire evacuation procedures
- Food storage procedures.

The function of document control or the computerisation of information is to ensure that there is clarity of information for the people using it and that everyone is singing from the same hymn sheet.

Oversight

A system of oversight is required to ensure that the written procedures are applied evenly at all times. The system selected needs to be matched to the requirement but, for example, having Quality Assurance Auditors work across all the shifts is a good method in spotting issues where methods are drifting apart. An auditor will be able to see differences occurring and can raise the issues in their report or indeed take the issue to the shift management straight away for it to be corrected.

Senior management need to be aware of the need for consistency and ensure that any differences between shifts are corrected. If two shifts have a different procedure for booking in raw material deliveries, there is a need to decide which the better method is and ensure that method becomes standard across all shifts.

Listening

If there are issues on the shop-floor with inconsistency in *how* things are done, then people will spot them and inevitably people will talk about it. This talk can sometimes be misinterpreted as moaning, but if a positive view is taken, it can be seen as an improvement opportunity. All you have to do is listen.

In summary, the issues around inconsistency between shifts can be resolved with a level of control, some oversight and simply listening to the people on the shop-floor. Using human nature as your detective is a very efficient way of finding improvement opportunities, but you have to listen to, and feel what, people are saying.

Improving communications

Improving *how* things are done often comes down to improving coordination between people, to ensure that everyone has the opportunity to contribute to the team effort. The first stage in any team building is effective communication. A good place to look for improving *how* things are done is to look at the area of communications.

Communications need to be planned and designed. Without an effective communication plan, there are no guarantees that the correct messages will be sent or received.

In a food production environment, it is especially important that communication is good and for this reason various systems have been developed over the years.

Communications systems have evolved from early beginnings. An example is the communication between the production department and the packing department in a high-risk food factory. The two departments are linked by the conveyor that takes the sealed tubs, bags or trays of food from the high-risk area to the low-risk packing department, where they are placed into cardboard boxes for transport to the retailers.

If there is a problem with the date code on the packs, this will often be spotted in the packing department. But how can this issue be communicated quickly to the high-risk team (see Figure 8.1)?

Question

- How is this situation handled in your factory?

How can the packing team communicate with the team on the other side of the wall?

- Banging on the wall?
- Shouting through a hatch?
- Just stopping the conveyor?
- Switching on a light or sounding a buzzer?
- Using an intercom system?

Quite often in the design of a food factory, communication is not considered. How does the warehouse supervisor get hold of the shift manager? How does the machine operator

Figure 8.1   How does this filling machine operator know if there is a problem in the packing department? Communication needs to be planned to ensure that messages are both sent and received correctly.

call for an engineer? How does the shift manager get hold of the technical manager, at 03:00 on a Sunday morning?

Communications are important, but in many food factories they have not been planned or designed and so are using methods that are the equivalent of banging on the wall.

What systems of communication should be in place?

Questions

- Take a moment to describe the systems currently used in your factory and then describe three ways in which communication could be improved.
- Remember, we are talking about *how* we communicate here, not *what* we communicate.

The future?

In this era of communication, the use of mobile voice systems is widespread; walky-talkies, mobile phones, DECT phones, tannoy systems, pagers, are all common place.

How about the machine operator being able to send an email to an engineer from the control panel of the machine? Or leaving a note for the operator on the next shift right there on the control panel (see Figure 8.2).

How about standard methods being held on a central computer system? When a change is made, all people trained in that system are sent a message and are invited to a session to be retrained in the new method.

A quality check should be made every 15 minutes. It has been 17 minutes since the last check. The computer control system stops production until the check is complete and the

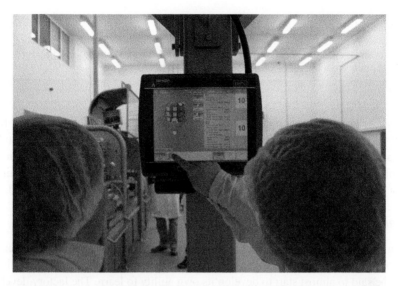

Figure 8.2 Modern control systems allow for a machine to be linked to the computer network. Send emails from the shop-floor or even see the training video for the change in SOP that was developed on the night-shift.

factory manager is sent a message to say that a test was carried out late. I think that next time the check will be on time!

Modern communication systems mean that if enough thought is put in, then communications can be planned and enforced. However, the needs for other methods of communication do not disappear, because it is through effective communications that motivation and enthusiasm can be encouraged. Creative new ideas can come about that would otherwise be missed.

The rules of communication are that it must be routine and regular, two-way and not part of a blame culture. Returning to discipline, start of shift meetings, end of shift meetings, waste meetings, performance meetings, the list goes on. The key here is that the meetings are seen as *how* things get done. Meetings need to be effective, which means that they have to make a difference to the business. Meetings should be short and conducted in a business-like way.

There are some occasions where effective communication is required, which are not routine and regular. Planned communication in the event of an emergency situation can often save money and, sometimes, even lives.

Questions

Just jot down the communications plan that your factory has in the case of a fire or emergency evacuation.

I guess most of you will have a fully detailed plan, with roles allocated to engineers, shift managers, security personnel; perhaps you even have hygiene crews prepared to allow the quick reoccupation of the factory in the event of a false alarm.

Questions

Now jot down the communication plan in the event of a major breakdown in your factory, which will put a production line out of action for a few hours.

This last point is probably more difficult to recall. Maybe your factory does not have a plan for such an event and you play it by ear each time it happens. Would it be a good idea to have a communication plan for some major events? A breakdown, a raw material supply problem, a hygiene problem. Sometimes this is called a crisis plan, but other times it does not exist at all and the communications after such an event are never planned.

The learning factory

This is a term sometimes given to a factory where continuous improvement in performance is occurring. The factory systems and communication are so effective that the factory can be said to almost start to develop its own ability to learn. The factory develops a base of knowledge that improves its performance, and people within the factory are seeing the benefits of working in a collaborative rather than a competitive way.

If you burn a finger on your right hand, you learn that it hurts but you also apply that learning to your left hand, and your arm, and your foot. In a learning factory, developments and improvements made in one department or on one shift are adapted and applied to other situations. Waste reduction on one machine is applied to all machines, and productivity improvements on one shift are applied to all shifts.

In a learning factory, the aim is not to be the best performing shift. The aim is to be the shift that is seen as the source of the best ideas, the best disciplines and the best teamwork. In a learning factory, the systems in place are designed to ensure that good practice is spread, not kept on one shift. The factory takes on a sharing and supportive culture rather than a competitive one. The whole factory performance increases as a result of the universal application of good ideas and disciplines.

Questions

• When was the last time that you shared an idea with someone in another department or shift, which would help improve their performance but have no effect on your own?
• Taking it one step further, when was the last time you shared an idea that improved their performance (and that of the factory as a whole), but reduced your own?

In a learning factory, the performance of the whole business is important. The performance of a department is seen as secondary (see Figure 8.3).

Getting the culture of a factory, with its entire people, machines and materials, and all of its systems and methods to a point where the factory learns, is beyond the scope of this book. Truly high performing factories around the world are moving towards this aim, in order to stay ahead of their competition.

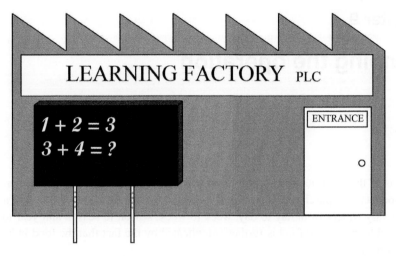

Figure 8.3   A learning factory is one that behaves in a way that optimises performance and is truly lean in its way of operation.

Summary – Improving how we do things

This chapter looked at how things are done in food factories. Because Lean Manufacturing is centred on a process of continuous improvement, that can also mean continuous change. It focused on important areas involved in change management. Uppermost in the process of improvement and change is the topic of communications. It was seen how poor communications can reduce the rate of change or even send the improvements being made into reverse. Communication is vital to the whole subject of Lean Manufacturing and must be functioning at a high level if continuous improvements are to be introduced successfully.

The end of the chapter briefly looked at the concept of a learning factory. It must be remembered that Lean Manufacturing is a bit like a road map. It allows you to plan a journey towards world-class performance, but the route you take and how long the journey takes is up to you. The chosen route must be right for you and your factory and you may have to do a few U-turns if the route is not working. But if you have strong communications with the whole workforce of the business, then the journey will be a bit easier to carry out.

Chapter 9

Planning the operation

All areas of the food manufacturing industry are in business to satisfy consumers with their products. The purchasing behaviours of food consumers are very difficult to predict and can vary widely from day to day, week to week, season to season and year to year. This variable market for food is further complicated by the fact that the food industry is very competitive.

Before looking at the issues around planning for a food business, it is important to understand the kind of things that have an impact on the market and the purchasing behaviours of the consumers.

Food consumption and purchasing decisions are influenced by consumers' wants and needs, the same as with all goods and services.

Consumer decision-making

So, I need food for my breakfast tomorrow – I'll buy bacon and eggs. That sounds like good news for bacon and egg companies – but wait.

Here are the sorts of things that go through every shopper's head when making a purchasing decision:

What kind of bacon? – Should I buy streaky, smoked back or this dry cure? Eggs – should it be barn eggs or free range? – Oh, and how about these ones with extra omega 3 oils – that's good for me, I heard it on the TV. Should I buy the branded products or the retailer own brand? How big a pack should I buy, what is the shelf-life? Oh perhaps I should avoid bacon and eggs altogether and buy breakfast cereal and some milk or maybe some bread for toast. Oh, those crumpets look nice but it's nearly Easter, so perhaps some hot cross buns…

Not every shopper has these thought processes for every food purchase, but by buying the same products regularly, it makes the job of planning a little easier. However, there are significant numbers of shoppers who do change their purchases on a regular basis in response to special occasions, a change in the weather, new product launches and new packaging designs or, perhaps most significantly, price promotions.

Handbook of Lean Manufacturing in the Food Industry, First Edition. Michael Dudbridge.
© 2011 Blackwell Publishing Ltd. Published 2011 by Blackwell Publishing Ltd.

Figure 9.1   The shopping habits in the retail stores will have an impact on the type and quantity of food made in your factory. The changes in sales patterns will feed through ever more quickly as the information systems of the retailers allow them to react to changes in the marketplace.

The weather will be good this weekend – I'll buy food for a picnic.
There is football on TV tonight – I'll buy a pizza.
I need sliced bread – which brand is on BOGOF?
Oh! That new ready meal looks good. I'll buy that instead of the usual meat pie for Wednesday (see Figure 9.1).

Questions

- Have a think about the products made in your factory and how the behaviours of the consumer have an impact on your business in terms of the work of the factory.
- What was the impact of the last big price promotion that you did?
- What happened to the sales of your products when a competitor launched a new product?
- What was the result, in your factory, of an advertising campaign?

Consumer need is constantly changing. By understanding the market in which the food industry operates, it is time to start thinking about how a factory can develop a plan for production so that this food can be made efficiently.

Planning systems

Factories need to plan ahead to ensure that all the elements of production are in place for the factory to work. The basic elements of production are:

- **Raw materials** – the ingredients for the foods to be manufactured.
- **Packaging materials** – not just the food contact materials but also the boxes and pallets, the labels, the printing ink, the stretch wrap, maybe the plastic crates.
- **The machines** – the mechanical parts of the factory required to make the products. The mixers, the cookers, the coolers and freezers, the packing machines, the forklift trucks.
- **The people** – most factories require skilled and unskilled labour, managers and supervisors, engineers.
- **The space** – there needs to be a space in which the production occurs. In some factories this need to be of a special high risk design. It is not possible to make or store food on the car park at the back of the site! Planning space is a strategic process, so the efficient use of space is an important principle of Lean Manufacturing.

There is a need for enough of these elements of production to satisfy the demand for our products, but not too much as that would generate inefficiency and waste, or too little as that would prevent us from giving the consumers what they want and need.

The major elements of production have to be planned to ensure that there is a sufficiency of each element to just meet that demand. If the planning is good, there should be a chance of making a profit. If the planning is poor, the profit will be reduced or maybe even become a loss.

Planning systems are vital in a modern food business and are often placed at the very centre of that business to act as the hub of the operation. The planning allows the business to make all the necessary decisions about the core of the business, the manufacture of food. Some businesses have their management and communication structures and systems designed so that the job of planning is central to all activities (see Figure 9.2).

Below are some typical planning systems showing how they work and control the five basic elements of production.

Planning horizons

First, the process of making a plan is an exercise aimed at improving decisions within the business. This is an important first step in thinking about planning. It is possible to have a plan for the next hour, the next shift, the next week or the next year.

Decisions on some of the elements of production need to be made well in advance. If launching a new product in six weeks' time, there may be a need for more machines, or even more trained staff. If planning has gone well, then the machines will be in place at the right time with staff trained to run them. If the business is to grow by say 50% in a year's time, you may need to build more space or need to adjust shift patterns.

The main area of planning to consider is the short-term stuff focused on the manufacturing operation. The best way to look at this is to think through the elements of production and make sure that the plan helps decisions in all areas.

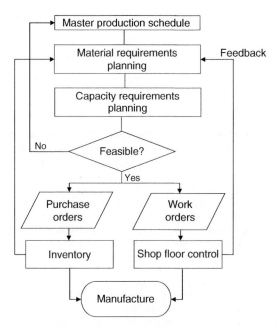

Figure 9.2   Short-term planning systems all start with some kind of master schedule; once that plan is decided, materials and factory time can be planned with a works order/production plan issued for execution. In the food industry, all short-term planning is often carried out in the space of a few minutes when a surprise order arrives from your customer. Computerised systems called ERP systems carry out most of the work, with the support of someone to oversee it.

The need to forecast the sales

Immediately there is a need to know how much of your products are going to be sold. Hopefully, your customer will have placed the order for your products well in advance, so it is known exactly what is required. More often though, the customer will not tell you how much they want until the last possible moment, so the whole of the planning process is based on forecasts of the likely sales.

The planning and systems necessary to get the ingredients needed to make the products required are:

Food materials planning systems

Food materials come in all types. They can be long shelf-life, frozen or dried or canned, or shorter life, fresh vegetables or dairy products, meat or fruit. The planning system required to ensure that the correct quantity is available for production is always the same.

First, how much is needed? The planning system will have to be able to calculate the quantity of material required to meet the order, or more often the forecast of the order. This process is nearly always carried out using a system called a Bill of Materials (or BOM). The BOM is a recipe of the product to be made. The BOM contains the quantities of materials required to manufacture each product, but will also contain an allowance for waste and yield of that material in your production process.

Case study

Let's look at a BOM for a simple cheese sandwich. The ingredients required are:

- White sliced bread
- Butter
- Blocks of Cheddar cheese.

The bread is delivered in loaves, with 16 slices per loaf. We require 2 slices per sandwich pack. Remember that each loaf comes with two crusts that are discarded, so an 800 g loaf will have a yield of only 88.8%. All other faulty or damaged slices are returned to the supplier to be replaced, so there are no other losses on the sliced bread.

The butter is delivered in 10 kg blocks, is softened using a warming process and is then placed in a mixer to improve the spreading. When we unwrap the butter, some of it sticks to the plastic wrapping material and when we mix the butter we always leave some in the mixing bowl. This means that for every 10 kg of butter we purchase, we only get 9.8 kg onto the production line. That is a 2% yield loss. We need 20 g of butter per sandwich, but on most days the average used is 22 g. That is a 10% giveaway.

The Cheddar cheese is delivered in 10 kg blocks and is grated before being put onto the sandwich. There is a 2% yield loss on the grating machine, because it keeps blocking and needs to be cleaned out. There is also a 5% giveaway on the cheese when it is put into the sandwich. The target weight for the cheese is 50 g.

The BOM for our Cheddar cheese sandwich is shown in Figure 9.3.

If we want to make one sandwich, we have to provide 99 g of bread, 23 g of butter and 54 g of Cheddar cheese at the start of the process. The order for tomorrow is for 50,000 sandwiches.

Figure 9.4 shows our food materials plan for tomorrow.

We need 4955 kg of bread, 1136 kg of butter and 2688 kg of Cheddar cheese to meet the order of 50,000 sandwiches.

BOM for Cheddar Cheese Sandwich			
Material	Target Weight	Yield %	Actual Weight Required
White Sliced Bread	0.088	88.8	0.099
Butter	0.02	88	0.023
Cheddar Cheese	0.05	93	0.054
Total Weight	**0.158**		**0.176**

Figure 9.3   A BOM for a simple cheese sandwich.

BOM for Cheddar Cheese Sandwich				
Material	Target Weight	Yield %	Actual Weight Required	Required for tomorrow (kg)
White Sliced Bread	0.088	88.8	0.099	4955
Butter	0.02	88	0.023	1136
Cheddar Cheese	0.05	93	0.054	2688
Total Weight	**0.158**		**0.176**	

Figure 9.4   A BOM with the order for 50,000 sandwiches added.

Figure 9.5 The market for pre-packaged sandwiches is very large and a difficult market to supply. The short shelf-life and fluctuating demand add to make the planning of raw materials for the factories very complex.

It gets a bit more complicated than that

Besides the BOM for cheese sandwiches, it could be that our factory also makes other sandwiches and that the food materials are sometimes the same. There will be a BOM for chicken mayo sandwiches that uses the same bread. The Cheddar cheese is also used in the cheese and pickle sandwiches. All BOMs for all products have to be calculated and the total quantity of food materials needs to be added up (see Figure 9.5).

In a typical factory, this can be a huge task and as a result the calculations are carried out using computer programs. Software can be written using spreadsheets to make the calculations, but some companies use Material Requirements Planning (MRP) to carry out this task.

MRP systems are computer programs that help with all the tasks associated with raw material supply to a factory. Such a program looks after the BOMs and recipes; it contains all of the information about the shelf-life of the raw materials, expected yields and allowed wastage levels. The final section of an MRP system is the stock of material that is held in the factory. The stock held is the final part of the materials planning system. By using the BOM, the quantity of material required has been calculated, but how much should be ordered from our supplier?

Our example shows the bread that needs to be supplied fresh each day; there is no opportunity for the factory to carry any stock of this item, unless an emergency stock is held in a freezer, just in case!. Butter and cheese have a shelf-life that would allow the factory to carry a stock and so buy in larger quantities and so maybe get a lower price. If the factory has storage space, it can be an advantage to hold a small stock of these items. This can reduce the number of orders placed, deliveries off loaded and invoices processed.

However, there are disadvantages to running a factory with high levels of stock. Remember JIT from earlier in this book.

If the customer decides that they want a low fat spread on their product rather than butter, it can leave your factory with a large stock of raw material that it cannot use. This can be very costly. One of the principles of Lean Manufacturing is to run the business with as low a stock level as possible, so as to minimise the costs associated with storage. This area is discussed in other parts of this book.

Questions

- Have a quick think about the last time you had an issue in your factory where you ran out of something that you needed.
- What was the cause of the problem?
- How did you get around the issue?

Back to our Cheddar cheese sandwich

Let us assume that a decision has been made to keep a small quantity of stock of butter and Cheddar cheese, maybe about one day's stock of each. That would be around 1000 kg of butter and 2000 kg of cheese. This stock needs to be counted to check that it is still there; the factory could have had a worse than expected yield or a very high wastage, and then the actual stock is used to calculate how much cheese and butter needs to be ordered (see Figure 9.6).

Now there is a materials planning system that will ensure sufficient materials to carry out the required production.

In some factories, especially those making short shelf-life foods, the actual order for your products is placed very late. In these circumstances, the whole of the materials plan is based on forecasts of likely orders. The accuracy of these forecasts is generally not good and as a result the factory can easily find itself with either too much or too little material to make the required product. Techniques have been developed in these factories to ensure that errors in the forecasts have only the minimal effect on the costs of manufacture. These techniques are discussed elsewhere in this book.

The next part of the plan is the element of production, packaging materials. This can be a lot simpler than food materials, but there are still some areas that need to be thought through.

First, packaging materials have a long shelf-life, which means that in holding stock there is less of a risk of the material going off or becoming unusable. Packaging materials can be very product specific, so there is still a danger of packaging becoming redundant following a change of recipe or a product being delisted. Printed and non-standard packaging materials can take a long time to produce at your packaging supplier's factory and is much cheaper if ordered in larger quantities. As a result, often up to three months' supply of packaging materials could be held by a food factory. Planning for packaging supply is carried out in a similar way to that seen above for food materials. Stock checks are carried out, sales levels are forecast and a decision is made when an order for more packaging needs to be placed.

MRP systems are employed in the planning of packaging supply in the same way they are used for food materials (see Figure 9.7).

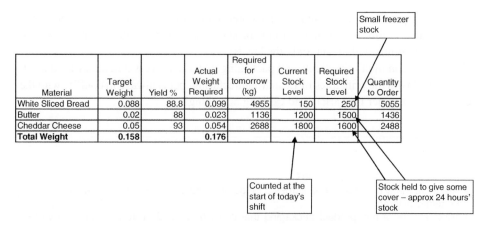

Material	Target Weight	Yield %	Actual Weight Required	Required for tomorrow (kg)	Current Stock Level	Required Stock Level	Quantity to Order
White Sliced Bread	0.088	88.8	0.099	4955	150	250	5055
Butter	0.02	88	0.023	1136	1200	1500	1436
Cheddar Cheese	0.05	93	0.054	2688	1800	1600	2488
Total Weight	**0.158**		**0.176**				

Small freezer stock

Counted at the start of today's shift

Stock held to give some cover – approx 24 hours' stock

Figure 9.6   Taking stock into account in the calculation of material requirements.

Figure 9.7   Stocks of packaging can be held to allow the purchase in large quantity to drive the cost per pack down. However, there is still a risk that the packaging could become redundant as food fashion causes changes to recipe or the product may be delisted altogether.

The risks associated with packaging planning can be greatly reduced if packaging is designed to be similar across a range of products. There are some good examples of this in the food industry, where companies have designed the packaging for a range of products with as many common items as possible. In our sandwich factory, all sandwiches are packed in the same skillet, with product differentiation being given using a sticky label.

All sandwich packs are placed in the same outer with a different out-case label being the only variation. In this way, a change to the product range does not immediately result in packaging becoming redundant, only the labels!

The use of common ingredients and packaging materials across a range of food products fits well with the principles of Lean Manufacturing, but is always in conflict with the need of the consumer for variety and exciting new things.

Questions

- How is your packaging designed?
- Are there many common items across a range of products or does each product seem to be totally different.
- Make a note of any product packaging that could be looked at, to try and improve its uniformity.

Next on our list of the elements of production are the machines used in the production of the foods. The planning of machines is a key step in the planning of a food factory. Often it is the machines that are the bottleneck in production. A bottleneck is a restriction in speed and capacity. A mixer can make 4 tonnes of dough per hour, but the oven is only capable of taking 3 tonnes per hour. The oven is said to be the bottleneck of the process. Earlier it was seen that counting and placing of pepperoni on a pizza assembly line was the bottleneck in a manual operation.

The task of planning in a food production process is to ensure that the output of the bottleneck is maximised by scheduling the tasks to ensure that the bottleneck process never occurs (see Figure 9.8).

In order to schedule and plan a production process, the first requirement is for knowledge of the speed of the bottleneck process. If the process is planned to run any faster than the bottleneck, it will not be efficient and the product will not be well controlled.

Questions

- Have a think about your production process; where is your bottleneck?
- When everything is running well, which part of your process stops you from going any faster?
- In a well-designed process that is well balanced, it can be difficult to spot the bottleneck, but there will be one.

Once you have identified the bottleneck, what is the maximum speed that can be achieved by that section of the process? That maximum speed should be investigated and checked to ensure that the number is as good as it can be, because the whole production plan will be based on that one number. If the production plan is based on any other speed, then the plan will not be optimal and will only be achieved by carrying out work that is not adding any value to the product and that is not part of Lean Manufacturing (see Figure 9.9).

Figure 9.8   The bottleneck process in this meat slicing operation could be any stage of the process. From feeding the meat in at one end, packing, labelling or even stacking the pallets.

Figure 9.9   Team members looking at the tray-sealing bottleneck on the line to ensure it is running at its best. Everything else on the production line is capable of running faster than this tray sealer, which has a maximum output of 200 trays per minute. It is vital that the bottleneck is running at optimum performance if the line is to meet its plan. Other equipment on the line could run less than perfectly and the output of the line would not be affected.

Case study

For example, let's assume that the tray sealing operation is the bottleneck in the production of a ready meal. Everything else on the line can go faster than the 50 trays per minute achieved on the tray sealer. We can weigh chicken at 80 per minute, we can deposit sauce at up to 70 per minute, and we can pack the trays into boxes at up to 70 per minute. If we fill the trays at the speed of 70 per minute, but only seal them at 50 per minute, we will make 20 trays per minute more than the tray sealer can cope with. We could decide to run the line like this and rack off the excess and feed them through the sealer during a product changeover or at the end of the shift. The Lean Manufacturing rules that are broken here are many.

We have to double handle the trays, we have to pay someone to rack the trays off and label them. We have to store the trays somewhere to await sealing. We have to count the trays racked off and add them to the ones sealed, to make sure we make the correct number before we changeover. We have lost control of the traceability of the product, as the trays are not sealed in the order they are made. The list goes on. By the time we have built in all hidden costs, the 20 extra trays made per minute are not making any money!

Lean Manufacturing principles would say that it is better to produce at the speed of the bottleneck, 50 per minute, avoid all hidden cost and control issues and as a result be able to plan the production more accurately. The time and effort of the people on the line should be spent on increasing the speed of the bottleneck to more closely match the potential speed of the process.

For example, the bottleneck on our sandwich operation is the sealing of the skillets once the sandwich is made. The maximum output of the line is 50 per minute. A plan for tomorrow might look like that shown in see Figure 9.10.

Short Interval Control (SIC)

The function of all planning is to aid decision-making, so the plan for line 5 is just adequate but could be better. One of the principles of Lean Manufacturing is called Short Interval Control (SIC). SIC is helpful on the shop-floor as it provides a regular check for the production teams about how they are performing. On our Line 5 plan, Cheddar cheese sandwiches are being made for over 16 hours to achieve our output of 50,000 packs. It would be useful to have a plan broken down into one hour or even half hour chunks, so that progress can be monitored on a regular basis. That way, decisions can be taken as early as possible if there is a delay in the plan. This kind of SIC would mean the plan might look like that in Figure 9.11.

In this plan, there is also a column to allow the actual time in which each section of work is completed. This will be useful, as it will become immediately apparent if the production is not going to plan.

SIC can be applied to all the factors of production, by providing a regular measure of performance, to help in decision-making. For example, it is known that 2838 kg of Cheddar cheese is needed to complete the order of 50,000 sandwiches. Would it not be

Production Schedule for Sandwich Line 5				
Product	Rate per minute	Order	Time Start	Time Finish
Cheddar Cheese	50	50000	6:00	22:40
Changeover		1	22:40	22:50
Cheese and Onion	50	8000	22:50	1:30
Changeover		1	1:30	1:40
Chicken Salad	50	4500	1:40	3:10
Clean Down		1	3:10	6:00

Figure 9.10 A sample production schedule.

Production Schedule for Sandwich Line 5				
Product	Rate per minute	Order	Time Start	Time Finish
Cheddar Cheese	50	3000	6:00	7:00
Cheddar Cheese	50	3000	7:00	8:00
Cheddar Cheese	50	3000	8:00	9:00
Cheddar Cheese	50	3000	9:00	10:00
Cheddar Cheese	50	3000	10:00	11:00
Cheddar Cheese	50	3000	11:00	12:00
Cheddar Cheese	50	3000	12:00	13:00
Cheddar Cheese	50	3000	13:00	14:00
Cheddar Cheese	50	3000	14:00	15:00
Cheddar Cheese	50	3000	15:00	16:00
Cheddar Cheese	50	3000	16:00	17:00
Cheddar Cheese	50	3000	17:00	18:00
Cheddar Cheese	50	3000	18:00	19:00
Cheddar Cheese	50	3000	19:00	20:00
Cheddar Cheese	50	3000	20:00	21:00
Cheddar Cheese	50	3000	21:00	22:00
Cheddar Cheese	50	2000	22:00	22:40
Changeover		1	22:40	22:50
Cheese and Onion	50	3000	22:50	23:50
Cheese and Onion	50	2000	23:50	0:50
Cheese and Onion	50	3000	0:50	1:30
Changeover		1	1:30	1:40
Chicken Salad	50	3000	1:40	2:40
Chicken Salad	50	1500	2:40	3:10
Clean Down		1	3:10	6:00

Figure 9.11 A production schedule with SIC.

useful to have an hourly check on cheese usage to act as a SIC on the yield and giveaway of this valuable ingredient?

There will be more on SIC in another section.

Manpower planning

The final element of production is the people required to operate the machines and work on the production line (see Figure 9.12).

Figure 9.12   Matching the available labour to the required output can be a complex area. Labour is an expensive element in the overall cost of food manufacture and needs to be planned with a great deal of precision if costs are to be controlled. Too much labour results in excess cost and low profit. Too little labour can result in manufacturing delays and poor service level to you customer.

Questions

- How is labour planned in your factory?
- How do you ensure that there are the correct number of people with the correct skills to run the machines and ensure that the quality and safety of the food being produced?
- Who carries out the labour planning?
- Could labour planning be improved?

Labour planning can be a complex area. There are numerous different agreements in place with factory staff, in terms of the degree of flexibility that they have in their contracts and what they can be asked to do in terms of job task and hours of work. Because of the unique character of the food industry in terms of its response to consumer demands there are many different arrangements in place to allow the business to respond, but at the same time being reasonable and fair to the staff in the production areas.

Questions

- What system of labour flexibility is in place in your factory?
- Is it seen as being fair and reasonable?
- How might the system be improved to allow the company to respond more readily to the demands of the market?

In some factories, labour is planned by the management on the shop-floor, in others the labour plan is the responsibility of a special part of the planning team.

Systems of labour planning

There are several systems used in the food industry to plan and control the use of labour in the production process. The aim of all the systems is to match as closely as possible the labour usage with the work that is needed.

Work planning

The first group of labour planning systems is really a work planning system. Labour is left broadly standard all the time and the quantity of work is varied to match the labour available. This is commonly used in the manufacture of long-life food products. The business serves its customers from a finished product stock and the work of the factory is to replenish the stock on a regular basis. Production lines can be shut down if the labour is not available and the lost product can be made later on. In this system it is possible to shut down a whole department for two weeks to allow holidays to be taken. Days of no production can be planned in to allow for maintenance or other activities to take place.

In food factories that provide short-life foods, such as chilled ready meals or fresh prepared fruit, the option of planning production around labour availability is not possible. Food has to be made in exactly the correct quantity on a daily basis and staff have to be made available to carry out the work.

Labour planning

This second group of labour planning requires a flexible approach from everyone, in order to closely match the labour available with the quantity of work to be done. The picture is further complicated by the fact that demand varies from day to day and the requirement is often not finalised until shortly before the work is done, or indeed sometime just after the work is done! There have been several systems developed to gain the kind of flexibility required.

Flexible contracts

This is where the people in production have an agreement with the company that allows for shifts to be of variable length; short-time working and compulsory overtime are a characteristic of this type of working arrangement. In order to provide security and fairness for the staff, this type of arrangement is sometimes built into a scheme of Annualised Hours, where the guarantee to the staff is for a certain number of hours per year in return for flexible hours (usually within limits) of work.

Overtime working

This is a system of flexibility where the shift is extended to allow the work required to be completed. The extra hours are usually paid at a higher rate, to encourage people to stay beyond their normal hours.

Use of agency labour

This system is widely used within the food industry, as it provides a great deal of flexibility for the factory in the use of unskilled or semi-skilled workers. Agency workers (sometimes called Contract Labour) are not employed by the factory. An agency is engaged to provide a number of staff between set times and are paid for providing that staff.

The attraction of this system is the total flexibility of the labour available. The factory pays for exactly the hours it requires and is not tied into any long-term commitment to the staff. The disadvantage of the system is that the temporary staff may be different each day and so generally can only be used on unskilled or semi-skilled tasks.

Questions

Before we move away from the area of labour planning, think about the systems that operate in your factory:

- Does the quantity of labour available always match exactly what is required?
- What do you do if you have too much or too little labour for the work you have to do?

The importance of sticking to the plan

After looking at how to put a plan together for the main elements of the food production process, this section deals with the disciplines of sticking to plan and why that is important in a factory that is trying to use Lean Manufacturing techniques.

The very essence of Lean Manufacturing is that slack is cut out of the factory systems and processes. In order to do this successfully, it is vital that discipline is in place. The capacity of the factory to deal with the unexpected is not present, that was in the slack that has been removed; in order to deal with unplanned events, costs will rise and that needs to be avoided if at all possible.

The best way to avoid unexpected events is to plan carefully and fully and to ensure that the plan is met. In a Lean factory the spare capacity, the spare people, the spare machines, the spare raw materials, the spare packaging and the spare space have all been removed. The factory will only be able to do its work by sticking to a strong plan and building in the flexibility and responsiveness that it needs without adding extra cost.

Short Interval Control (SIC)

It is very important in a system, which relies on planning for its success, to know as soon as possible if the plan is not met. This is achieved by a system of SICs that are used to monitor the important elements for the achievement of plan.

Systems are set up to capture data from the shop-floor and to use that data to compare against the planned position. If the SICs indicate that performance in an area is not what was planned, then decisions can be taken early to make some corrective actions and get back on plan as soon as possible.

Systems of data capture can be simple counts of the quantity made or can be measurements of the average weight of packs produced so far. They can be an hourly measure of waste generated or a score from an hourly taste panel. Any measurement that aids decision-making and helps the process to achieve its plan can be called an SIC.

Why is sticking to plan so important?

The production plan is usually issued by a person who has all the knowledge and experience of the production process, to ensure that the most efficient plan is put together. This will often be in conjunction with the factory management team. The plan will be used by all parts of the business to carry out their work.

The obvious users of the plan are the production areas, but the stores areas may well use the plan to organise their work in support of the production process (see Figure 9.13).

> Oh, I can see that cheese and onion sandwiches are not on the plan until 22:50, so I will move the onions from the store to the chiller at 21:30 tonight.

Imagine the chaos if the production supervisor decides to make the sandwiches early because some spare staff were available.

The goods in supervisor is booking in vehicles:

> Oh yes, I can see that chicken salad is not on line until 01:40, so it's fine to deliver those labels we need at midnight.

Perhaps the production supervisor does not know that the factory ran out of labels yesterday and has had to place a rush order this morning to make up the shortfall.

The scheduling of a line is important but so is the speed of operation, the yield, and the waste generated by all these factors of production can have an impact on the factory's ability to meet the plan.

If giveaway is too high, the factory may run out of cheese before the order is complete. If the sealing machine is causing a high level of waste, the output will be down as well as the yield. All of these factors need to be communicated to the planners, to ensure that corrective action can be taken where necessary.

Some factories have developed an escalation system of communication to ensure that relevant factors are communicated effectively and in a timely manner. The line supervisor will inform the shift manager if the line falls more that 15 minutes behind plan. The shift manager will inform the factory manager if the line falls 30 minutes behind plan.

There can also be escalation for waste, yield or any other factor of production that is part of the SIC system of control.

The aim of all the planning and control systems is to aid decision-making and ensure that factory performance remains at a high level. By using escalation systems, decisions can be made by the people who need to do so, without them making all of the decisions for everyone.

Figure 9.13   Teaching a vision system to count the production on a line will allow very accurate information to be used in the SICs. As well as counting, this system can also check for visual quality of the products, so an SIC for quality is also very easy to do.

Questions

- What is the system of escalation at your factory?
- Are there clear levels of reporting?
- How about late at night; is there a system in place?

How can we make the plan work?

How to ensure that everything required is completed:

1 The first assumption here is that the plan is correct. The plan is achievable and reflects the true position in terms of the speed of production and the availability of ingredients and packaging materials:

- If the plan is not achievable, then this needs to be raised as an issue as soon as possible, so that decisions can be taken. It could be that the machine is unable to run at the planned speed or that the labour or skills are not available to achieve the required performance or output. It could be that there has been a quality issue and a quantity of materials or final product is no longer available. It could be that there has been a breakdown on the line and time has been lost as a result.
- If the plan is not achievable for any reason, then a new plan needs to be made and communicated to ensure that everyone is kept fully informed of the decisions taken to correct the situation. A usual way that issues of this nature are raised is

Figure 9.14 A change of plan on this crisp production line. Everyone will need to be aware of the plan changes, so that there will be minimal impact on performance (picture courtesy of Ishida Europe).

at the start of shift meeting. The need for a new plan is raised there, so that everyone there is aware that the plan they follow currently will be updated soon (see Figure 9.14).

2 Once the plan is achievable, the discipline of sticking to that plan is the next step. It is vital that the whole production team works together to monitor the situation and try to look for issues that could delay or stop production. Who better to monitor the usage of cheese than the person who brings the cheese to the line? Who better to check the speed of production than the person stacking the cases onto a pallet? Who better to keep an eye on wastage than the person who is taking the black bags to the skip?

By allocating the SIC system to people in the team, it is possible to monitor performance of many factors and ensure that the target is met. The target is, of course, to hit the plan in terms of output, but to ensure that the standards of safety, quality and cost are also met (see Figure 9.15).

3 If the plan is correctly calculated, then it is important to remember that a minute lost is a minute lost. Sounds obvious but people often believe that time lost in production can be caught up later. "We will catch that two minutes up at the next changeover" is often heard. The question needs to be asked: "If the changeover could be achieved two minutes quicker, why don't we do it like that all the time and put the new quicker changeover into the plan (see Figure 9.16)?"

If the plan has been made on the basis of perfect performance, it is likely that any delays or issues will mean that the plan cannot be achieved. A plan should be built around a realistic expectation of performance rather than assuming that everything will be perfect. It is common for planners to use the average run rate from the past seven

Figure 9.15   Everyone in this production team could be monitoring an important aspect of production. The use of SICs gives quick feedback of performance to the team, so that corrective action can be taken before the issue gets out of control and becomes unrecoverable.

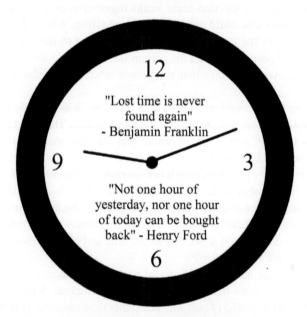

Figure 9.16   "Lost time is never found again" – Benjamin Franklin
"Not one hour of yesterday nor one hour of today can be brought back" – Henry Ford

Figure 9.17 Is this sufficient material to finish the production run? Will materials run out before the customer order has been completed (picture courtesy of Ishida Europe)?

days as a basis for a workable plan. The average yield over the last month is used to calculate raw material requirements. The average level of absenteeism is used for the labour planning. These numbers will produce a workable, achievable plan, but sometimes a plan built around a stretch target can have a motivational effect on the teams. If stretch targets are to be used, the team must agree that they believe they can be achieved.

4 One area where teams in food production often get caught out in their attempt to stick to the plan is in the area of counting. When materials are moving rapidly around a factory during food production, it can be very difficult to keep an accurate check on what is happening. "Is this the sixth pallet or the seventh?" This is especially the case during stocktaking activities, where small errors in counting can cause big issues because of a lack of materials or a lack of product. Counting errors can lead to lines stopped too early or too late for a changeover to the next product. These errors can have a very big impact on the production efficiency, especially if the error is not discovered for some time (see Figure 9.17).

Questions

- When was the last time that your team produced too much or too little?
- When did you last run out of a material that you thought you had enough of?
- When did your factory last have an issue with the accuracy of a stock count?

Summary – Planning the operation

After looking at the basic requirements of a planning process, it can be seen that in food factories the planning process is very quick, with orders arriving as late as possible for the company to react to and send the products to the customers.

The planning process operates to very short horizons and is often automated using ERP systems.

Making a plan work is all about a disciplined approach to production, the setting of an achievable plan and the constant monitoring against that plan. If production is falling behind plan, then the final way to make a plan work is to do something about it. Make decisions or get others to make decisions to bring the production back on plan. As a last resort, if corrective action cannot be taken, then change the plan and communicate the new plan.

A plan is simply an aid to decision-making. If the plan is not working, it will only encourage bad decisions to be made and as a result hidden costs will creep into the production process, and the principles of Lean Manufacture will leave the area very rapidly.

Adherence to a plan is a very common KPI in short shelf-life food factories, a measure to try and get a hand on the lever of the factors that are stopping a factory from sticking to the plan. These can be performance related or they can be related to the organisational skills of the factory teams.

Chapter 10

Start of shift meetings

As shown earlier, Lean Manufacturing techniques are concerned with improvement of performance, by the use of the efforts of a team of people from the business. An important part of the efforts to improve performance is communication between all the people involved. It is vital that everyone involved can see the results of their efforts, both good and bad, and the team can work on the issues of change together.

It is seen as good practice to hold a meeting with the team at, or close to, the start of each shift, to allow the team to come together and keep focused on the projects and improvements in hand at that time.

A start of shift meeting sets the agenda for that day for the team. It reviews the lasts shift's performance and tries to make today's shift better. Lessons are learnt from the things that did not go well and tasks are allocated to the team members to investigate, develop or install the improvement to ensure that issue does not arise again.

If you want today to be better than yesterday, you must do something different. The investment of 15 minutes of time can gain huge business benefits, especially in the fast moving world of food manufacture (see Figure 10.1).

Questions

- What is the system used in your factory at the start of a shift?
- Is there effective communication of the recent performance and the momentum to improve it?
- How do you know of any upcoming issues that might impact your performance today?

Monitoring performance and actions

The performance of a food production department is measured and monitored by keeping an eye on KPIs. Performance data is collected by the teams, or sometimes automatically during a shift, and this data is converted to a series of KPIs. These KPIs appear in the DWOR and are reviewed at the start of each shift in the start of shift meeting.

The measures that are reviewed are different for each business, depending on the circumstances at the time, but a few KPIs are commonly used in the food industry.

Handbook of Lean Manufacturing in the Food Industry, First Edition. Michael Dudbridge.
© 2011 Blackwell Publishing Ltd. Published 2011 by Blackwell Publishing Ltd.

Figure 10.1   A start of shift meeting ensures that everyone has a chance to contribute to improvements in the process and performance of the factory or department.

There will usually be a measure of the output. This could be the OEE, which is a measure of the quantity of product made compared to the quantity that could have been made if everything had run perfectly. OEE is discussed elsewhere in this book.

There will be a measure of productivity. This is the number of man hours used to make the product compared to the quantity that was expected to be used.

Waste will be measured and recorded on most DWOR and discussed at the start of shift meeting, as will issues such as time lost because of breakdowns of machines, accident statistics and quality measures.

From the discussion of yesterday's performance will come tasks or actions that need to be taken to prevent reoccurrence. These actions will be allocated to someone at the meeting to complete and as a result, the start of shift meeting is one of the main ways in which the performance of the factory improves. Actions will usually be recorded on an action board, by writing a brief description of the action along with who is carrying out the task and when the task will be completed.

The start of shift meeting reviews the actions that are current and everyone has a brief opportunity to let the team know about progress.

The final part of the start of shift meeting is to look forward in the shift that has just started and try to communicate any issues that are likely to impact performance. Are there any problems with raw materials or machinery, is staffing adequate, are there visitors during the shift or are new products being launched. Does anyone in the team know of any reason why the planned production for the shift might not be met?

This is a useful conclusion to the meeting and acts as a motivational spur to the team to achieve the plan and meet expectations (see Figure 10.2).

How should a start of shift meeting be organised? First, it is probable that a factory will have several start of shift meetings on a shift. One could be in the logistics and warehouse

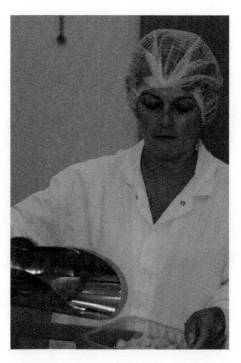

Figure 10.2   A start of shift meeting will allow the team to look at yesterday's performance in terms of their KPIs. They will also be able to report back on anything they were asked to investigate at yesterday's meeting. "Mary, can you have a look at the material yield during the last two hours. It was much better than the rest of the shift and there is a need to know why."

area, one could be in low risk production, one in high risk and one in the packing department. In a factory with long, continuous production processes, it is better to have a meeting for each line rather than each department. It is a good idea to limit the number of meetings as much as possible, to make sure that each area is well served but at the same time the meetings are coordinated so that the whole factory acts as one.

Where when and who?

Where?

The start of shift meeting should take place on the shop-floor in or near the production area. There are several reasons for this:

- The people who attend the meeting are team leaders, supervisors and managers, who work with the production teams. It is vital not to pull those people out of the factory if at all possible.
- Because the meeting is about improvement and change, it is important that everyone in the company can see that the process of improvement is going on at each and every shift.

Figure 10.3   Here the performance and action board has been placed in the very heart of the production area. The start of a shift meeting takes place here, not in an office or in the canteen with a cup of tea. The speed and efficiency is essential, as it needs to deliver improvements to the KPIs.

- The performance and action boards that form the centre of the discussions at the start of shift meetings remain in place on the shop-floor, so that everyone has an opportunity to look at the performance figures and see what actions are taking place to improve. Some businesses decide to make a feature of this information and create a performance centre in the production area. A performance centre is usually a meeting location that is shielded slightly from the noise of the shop-floor, but is openly visible to people in the factory (see Figure 10.3).
- The start of shift meeting needs to be short and focused. Typically, the meetings are completed in 15 minutes or less, with all the members of the meeting standing up. Because attendance at the meeting is compulsory, it is best to hold the meeting where the people are, rather than pull people out of the factory.
- Because the meetings occur every shift, having them organised to be quick and easy to attend makes the meeting more efficient. There is nothing worse than having a 15-minute meeting about improving performance that takes 30 minutes out of your shift every day.

When?

It is the start of shift meeting, so you can expect it to occur at the start of the shift. It is usual for the start of shift meeting to take place once the shift has got up and running, so typically an hour or so after the shift start time. That gives everyone a chance to coordinate the KPIs and to do some initial thinking about issues from the previous shift. Having a one-hour delay also allows for information to be gathered about potential issues on this shift that would hold back performance. It is vital that a regular time is set and there is a level of discipline about starting and finishing on time.

Who?

The people who attend the start of shift meeting are those who manage the activity during the shift. Typically, that would be team leaders and supervisors, with shift managers guiding the meeting. Also invited would be a maintenance-engineering supervisor, a quality assurance supervisor and maybe a member of the planning team. The team at the meeting are the people who will be driving improvement activity through the business and so it is important that everyone involved in that process is there, and no one is there who is not engaged in performance improvement.

Attendance is compulsory and the meeting occurs in a disciplined way at the same time every shift. If someone cannot attend the meeting, they will need to send someone else who has been fully briefed on the performance issues and on the actions that are carried out.

Performance and action boards

The start of shift meeting needs to be efficient and this is achieved in many factories by the use of performance and action boards.

The boards are usually the dry wipe type of white board with a grid marked on them to contain the required information. The performance board is usually a mimic of the DWOR and will contain the KPI data for the area. The action board is marked in a way that allows the recording of decisions of the meeting in the form of actions and tasks to be undertaken. Alongside the task is the name of the person who will get the task done and the date when the task will be complete.

The owner of a task or action is someone at the meeting and it is they who are responsible for getting the task completed. However, it is not necessarily his or her job to carry out the task, as that could be allocated to someone else outside of the meeting. However, the person named on the task is responsible for bringing the solution back to the meeting when the due date arrives.

What, who, when ...?

When a team is trying to improve by making many small changes to methods and systems, and the task of investigating and recommending changes is delegated to several different people in the team, it is important that a record is kept of what is being done, who is doing it and when the result is required.

Traditionally, the function of recording these types of issues was carried out by a secretary who would issue minutes of the meeting, so that everyone was clear about what had been discussed and decided by the team.

The administrative task of keeping track of much improvement activity would be costly and inefficient. Remember administration is a non-value adding activity, so Lean Manufacturing systems need to reduce it to a minimum.

High Risk Department Action Log

Action number	Task Description	Who?	When?	Complete?
1				
2				
3				
4				
5				
6				
7				
8				
9				

Figure 10.4   An example of an action log typically used to keep an eye on activity and the management of issues in the business. The action log does not have to be pretty. It needs to be a tool used to ensure that issues are solved and improvement is achieved.

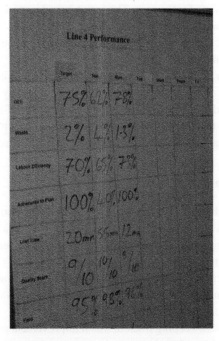

Figure 10.5   A typical performance and action board located on the shop-floor. The start of shift meeting will be held here.

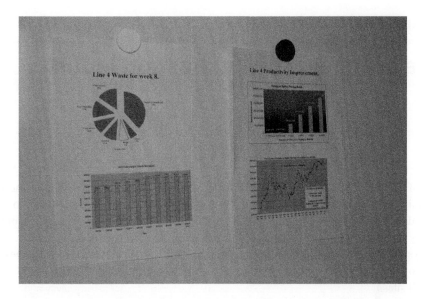

Figure 10.6   A typical performance and action board located on the shop-floor. The start of shift meeting will be held here.

The way that this work is carried out is usually in the form of an action log. This is a simple grid written on a white board in the production area to record the what, who and when (see Figures 10.4, 10.5 and 10.6).

Summary – Start of shift meetings

This chapter showed the importance of the start of shift meetings to the whole process of driving improvements in performance through a business. It is the team meeting at the start of each shift that carries out the vital roll of analysis of the KPIs from the previous shift and decides on areas for investigation and action.

The meeting also makes decisions based on the reports coming in from previous actions that have been reported back that day.

Finally, the start of shift meeting allows the team to predict today's performance and alert each other to things that may hold back performance.

A business that fails to call the team together at the start of each day is forcing the team to work as a group of individuals and is likely to cause people to work in isolation and on things that are important to them rather than the whole business. A daily start of shift meeting gives focus to the activity for the day and that focus is on performance improvement.

Chapter 11

The seven wastes in the food industry

One of the principles of Lean Manufacturing is getting things right first time and the minimisation and elimination of waste. Waste in a food factory is often easy to see. Wasted food and damaged packaging are easy to identify and the skips at the back of the factory are often full of waste that could, and should, have been avoided.

Physical waste of materials is only one type of waste. Such waste in a food business is easy to see in terms of wastage of materials; it appears in the waste bins and ends up in landfill sites.

The other forms of waste are more dangerous, as they are often unseen and persist for years before anyone notices.

If waste can be reduced, the money saved will stay within the business and will increase the profits of that business. In fact, for every pound saved in waste the profit will increase by one pound. That is a powerful tool for a business looking to improve its profitability.

The other types of waste that are more difficult to see, are no less costly to the business. Within Lean Manufacturing systems, there are seven types of waste that have been identified. A factory can be said to be operating to a high standard if all of these sources of waste have been minimised.

First, by looking at the seven types of waste given in the following examples, you can go into your factory and see if you can spot waste of all types being created (see Figure 11.1).

There are seven types of waste in a business

Motion

Operator motion

Travelling time for hands, people and physical distance between operations is a major cause of waste. For example, a person doing a job has to stretch or even walk to pick up the next tub of an ingredient they are using. The machine that grates the cheese is several metres away from the production line where the grated cheese is used. It is possible to reduce motion in a factory by using improved process layouts and applying ergonomics

Handbook of Lean Manufacturing in the Food Industry, First Edition. Michael Dudbridge.
© 2011 Blackwell Publishing Ltd. Published 2011 by Blackwell Publishing Ltd.

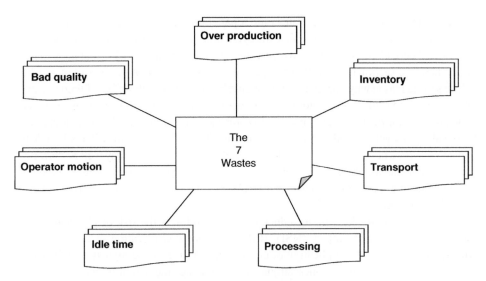

Figure 11.1   The seven wastes that occur in all operations.

Figure 11.2   Stacking crates onto a pallet, an example of a common task in food factories that is often improved and reduced in cost by a consideration of one of the seven wastes, operator motion.

to the workstations. It is also possible to ensure that everything is where it is needed, by using some of the techniques used in the 5S system of factory and work organisation.

Operator motion has three effects in a high speed and repetitive manufacturing task. First, each of the motions can take a fraction of a second longer to complete (see Figure 11.2).

An operative who is stacking crates onto a pallet on the end of a production line will sometimes struggle to keep up, because the cycle time for the task is longer than the available time. This can often result in two operatives having to be allocated to that task. In this book, there is a section on Line Balancing, which looks in detail at this aspect of a task.

The second reason why operator motion should be minimised is the effect of tiredness and fatigue on the operator. An operator in a food factory would, typically, be working an eight-hour shift with maybe one hour of breaks. Towards the end of a shift, the level of fatigue will have built up. Judgement and motivation will be impaired and the operator will not be able to maintain the pace of work. As a result of this end of shift fatigue, there may well be performance issues on the line and costs will increase as a result of operator errors or reject product.

By minimising operator movements in the tasks, it will be possible to improve performance and reduce fatigue.

The final reason for looking to reduce operator motion is the risk of physical injury through repetitive strain injury (RSI). RSI is a major consideration in the design of workstations through the application of ergonomics. A correctly designed area for work will reduce job cycle times, reduce fatigue and also reduce the risk of injury.

Waiting time

The time between finishing one operation and starting the next is non-productive and generates a large quantity of waste. All of the time that no product is being made is wasted time. Even worse, the non-productive time is also a waste of all the resources that have to be paid for. The power for the lights is wasted, the wages and salaries of the workers and managers are wasted, and even the cost of the phone system is wasted. All of those resources are there to support production; without production their cost is a waste.

There are a huge number of reasons why a production line may be idle and not producing product. These include:

- **Changeovers** – the machine is adjusted to manufacture a different product. This involves going from 200 g packs of rich tea biscuits to 300 g packs and adjusting the machine and changing the packing film.
- **Hygiene** – obviously a big feature in food factories is that a machine has to be cleaned in order to maintain product safety.
- **Maintenance** – the machine is having a routine maintenance procedure carried out.
- **Breakdown** – the production line has stopped because of an equipment failure.
- **Lack of materials** – the line may be waiting for more materials. This could be packaging film, trays or food materials.
- **Warming up** – it is a feature of heat sealing machines that they often require time for the heating elements to warm up after the machine has been switched on. This time can be a significant addition in time if the machine has been turned off for changeover, maintenance, breakdown or hygiene.
- **Waiting in the machine cycle** – as machines operate, there is often a complex series of events that go to make up one machine cycle. The control of a machine will be such that each step in the process will wait for the previous step to be complete. The small delays in the steps of a complete cycle can be measured in milliseconds and often, especially on high-speed machines, can be unnoticeable as a delay.

Figure 11.3   Waiting time has to be minimised in order to reduce the waste that it causes in food production. Waiting time has a variety of causes that will need measuring so that corrective action can be taken (picture courtesy of Ishida Europe).

A few milliseconds of unnecessary waiting time on a machine may sound insignificant, but it should be remembered that delay is occurring on every cycle of the machine.

For example, a machine has 5 milliseconds of unnecessary delay in a cycle time of 0.5 seconds (i.e. bagging frozen peas at 120 bags per minute). So, 5 milliseconds every 0.5 seconds. That is 600 milliseconds per minute. That is 36 seconds per hour. On our line, the output could be increased by 72 packs per hour, which is an increase of 1%, simply by looking at the delays in the machine cycle (see Figure 11.3).

All waiting time is waste and the cost of this waste can be very high, especially if the time has to be regained later through the use of premium rates of pay for the operators.

Overproduction

Too much of an item is produced "just in case". The plan requires 100 boxes to be made but there needs to be enough just in case one or two get rejected. Or the feed end of the line can be stopped to let the product flow through to the packing area if it is discovered that there are five too many made. Often overproduction is justified because the excess can "go into stock" and be used against tomorrow's order. This may well be the case, but the overproduction has caused costs to rise, even if the stock can be used. The excess stock has to be labelled and entered onto the stock records. It has to be counted and stored in the warehouse. Tomorrow it has to be counted again and the number communicated to the planning department. All of these extra tasks have a cost associated with them, so this is the waste associated with overproduction.

Obviously, if the overproduction ends up unsold or is sent to the staff shop, then the cost of overproduction rises steeply. Overproduction in the food industry is not uncommon and can be seen in company shops up and down the country. Overproduction of short shelf-life products is common. Retailers place their orders as late as possible for rapid delivery to their distribution centres. Food factories do not have the capacity to sit and wait for the order and so are forced to manufacture to a forecast of the order. Sometimes the forecast is very poor and as a result the food factory is left with stock in the despatch warehouse when the order has been fulfilled. It is hoped that the finished stock can be sold at full price against the next day's order, but often this is not the case. The finished good disposals to secondary customers are difficult to achieve and so the excess production is either sent to landfill or recycled.

The reasons for overproduction may have their solution inside the factory, with better counting and stock controls, or may have a solution in the wider supply chain with better forecasting. Either way, overproduction is a form of waste that needs to be minimised.

Overproduction of MG Rover cars in the UK caused large additional costs to the company in terms of storage and management. As times got tough, the company was forced to sell cars at discount prices to bring some cash into the company. Some companies who make long shelf-life foods were managed in a similar way, until it was realised that overproduction was wasteful. Most manufacturing companies will now manufacture as close to the time of sale as possible, to minimise stock holding and only make a product when they are convinced they will sell it.

Optimisation losses

Where a line is not in balance and one part of the line is capable of doing more, it cannot because of the machines feeding it or taking the product away. A fast machine is run below its capacity because of the systems around it. A wrapping machine is capable of 200 packs per minute, but the filling machine before it in the process is only capable of 150 packs per minute.

Optimisation losses are a feature of all food production lines. There is a need to minimise the waste in this area in the design and operation of the line. Looking for a bottleneck on the line and improving its output will reduce the optimisation losses of the rest of the line and speed up the process without significant investment. These no cost and low cost solutions are a feature of Lean Manufacturing.

The size of waste caused by optimisation losses can be difficult to detect, because the true performance of a section of a production line is held back. Referral to machine specifications and a chat with the original equipment suppliers may reveal potential performance in your production line (see Figure 11.4).

Defects

Where rejected products are manufactured, it causes waste material, extra effort in rework, unplanned material issues and late deliveries. When a product is made that has a critical quality fault, it forms a huge quantity of waste, some of which is obvious, but there is also hidden waste inside every rejected product. Not just the materials used to make and package the product, but the time it took to make, the effort and cost of waste

Figure 11.4 The output of this bread line was held back by the capacity of the cooling system. There was an optimisation waste in the ovens, mixers and all other sections of the line. Low cost modification of the coolers to improve performance resulted in an increased output from the line of 2%. There is still a level of waste due to optimisation loss in the line, but the production team knows where it is and has plans to get another 2% of it soon (picture courtesy of Ishida Europe).

disposal and the cost associated with a lost sale or late delivery. Defects are probably the worse kind of waste in a food factory. The need to "get it right first time" is paramount in a Lean Manufacturing food factory. The cost of having to repeat work is astronomical and unplanned.

It is vital in a Lean factory that quality is assured at each step of the process and that process controls ensure that the product reaching the end of the process is acceptable and meets specification. If during the process a critical fault is detected, the product must be rejected at that point to prevent the faulty product having any more value added to it.

Waste reduction in the food industry is a favourite topic for projects, as everyone realises that food materials in a skip at the back of the factory are very costly. A fully manufactured and packaged food product has an average sales value of £2000 per tonne. A full skip of 16 tonnes leaving your site can have a value of £32,000 and everything must be done to reduce the physical waste. A reduction in rejected product has an added incentive to those carrying out waste reduction projects. Every pound saved in a project is a pound of extra profit that day. It is an obvious area where manufacturing performance has an instant impact on the costs of the business. A food factory that runs at a lower level of waste than its competitors will be in a very strong position (see Figure 11.5).

Figure 11.5   Waste generated by the rejection of products that did not meet specification. Different colour waste bags were used to distinguish waste from different area of the factory. Different bags allowed the level of rejection to be established for each area. Improvement teams could then investigate reasons for rejection and minimise the quantity. The work also included analysis to see if a rejection was occurring at the earliest possible opportunity in the process where it would have its lowest possible value.

Inventory

Overproduction is a source of waste. Inventory is in a related area and is concerned with the stocks that can build up throughout a food factory. Inventory can be raw material and packaging stock and it can also be partly made product.

Using excess stocks to cover other shortfalls costs the business, in terms of the space required and the stock holding costs.

If a cake factory keeps running out of sugar, the size of stock held should be increased to make sure that is never runs out and a safety stock should be kept in a different warehouse just in case it gets a bit tight. In a factory making long shelf-life foods, the temptation is to make products to top up the warehouse and ensure that the customer is always satisfied from stocks held. Holding excessive stocks anywhere in the business, whether it is packets of biscuits in the despatch warehouse or spare printer cartridges in the stationary cupboard, takes up space and ties up money that could be used for other purposes. There is also the danger that the need for that item could disappear and that the item would be never required and have to be thrown away or sold at a big discount. Lean factories are run with minimum stocks in all sections of the business and other methods are found to prevent a lack of stock being a problem. For example, for raw materials, stocks can be kept low safely if the supplier of that material is able to respond quickly to demands from you for extra deliveries. Final product stocks can be reduced if your factory is able to produce every product every day without it having an impact on performance. The solution to inventory problems sits in the flexibility and responsiveness of your factory and its suppliers.

A lot of food factories have a method of production that is aimed at high levels of availability of materials or partly processed foods. This has built up over the years, so that the management of work in progress is a major task.

The waste within this entire inventory is difficult to see, but is there. Management of stock and work in progress is a costly task and if methods can be developed to reduce the inventory while maintaining performance, then the factory will be more efficient and lower cost.

This book has already mentioned the concept of JIT deliveries of materials from your suppliers, which is a good first step in the reduction of inventory. The next step is the reduction of work in progress inside the factory. No item is made until the latest possible moment before it is required. As soon as it is made, it is used. This gives a great advantage in terms of cost, but has a risk if the supplying machine has a breakdown.

As with all things in Lean Manufacturing, it is the journey towards no inventory that is the important feature. There will be parts of your process that need stock in order to protect the overall performance of the factory. The study of the seven wastes leads you in the direction of no inventory production methods, but maybe the best you can achieve is to reduce inventory.

For example, a factory moved to a JIT system with its supplier, so that packaging was delivered straight to the production floor every eight hours. There were three deliveries a day, seven days a week. The company needed less forklift drivers and forklift trucks. Earlier in this book, when examining VSM, it was shown that storage of materials does not increase their value in the eyes of the consumer. An empty warehouse is evidence that the waste caused by storage can be reduced. The business is now looking at expanding its production operations in their old, costly warehouse area, by upgrading and converting it to food factory standards.

Excessive transportation

This is the unnecessary movement of people, materials or information around a business. Waste of all types is created by such movement. Moving something does not add any value to it, it just adds cost. This rule applies equally to people, materials and information. The unnecessary movements are a waste of resources and need to be minimised.

When a food factory is originally designed, the flow of the product through the factory is thought about and the factory starts life with a very logical flow of materials and people. As already discussed, the food industry is constantly changing to meet the demands of customers and consumers and as a result modifications to the original design are often made. Within a short space of the time the movement of materials and people is no longer logical and is sometimes very wasteful. New walls are built, new machines are installed and new procedures and processes are introduced.

The costs associated with excessive transportation are clear. There will be costs associated with the movement in terms of energy. At its cheapest, this will be electrical energy to drive a conveyor and at its most expensive, it will be a person pulling a truck containing a pallet of material. In order to reduce the cost of transport in your factory, you need to look at the distance travelled and the cost. The simplest way of reducing the cost of transport is to reduce the distance moved. Redesigning your food factory will not be an easy task, but there will be local opportunities to reduce the distance travelled and maybe install a conveyor to reduce the cost.

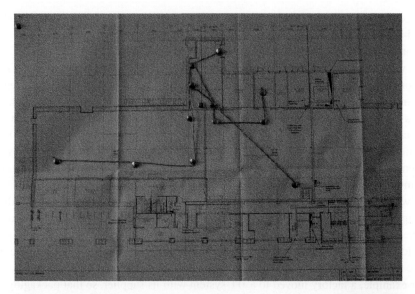

Figure 11.6   A simple string diagram can be used to plot the movement inside your factory. It will also let you plan the best location for machines and maybe get the factory working with less wasteful transportation.

A useful technique here is to make a string diagram to plot the movements around the factory. It will soon become apparent where there is a lot of movement and where improvements would yield the best result. As with many Lean techniques, it is often better to make improvements that have a big "bang for your buck", where you gain the maximum improvement from your work (see Figure 11.6).

Seeing the seven forms of waste

The seven forms of waste are in every business; the trick is being able to see the waste rather than accept "that is how it has always been". People can tend to become "blind" to even the most obvious forms of waste, so some of the forms that are more difficult to spot need some special techniques to bring them to the surface.

How can waste be spotted in all its forms and, therefore, how can an opportunity be pinpointed to reduce it or preferably eliminate it. How do you quantify the value of the waste and therefore prioritise projects to improve the situation.

Data and information is required for any decision that is other than "gut feel". The first thing to examine is information and data that proves that what is observed is actually costing the company money in terms of performance or physical waste. Successful businesses and successful managers do not go on "gut feel". Decisions must be based on real information. The larger the decision, the more information will be needed to support it. To predict the outcome of a decision, some measures must be in place before the change is made. The information must be:

Figure 11.7 One shift's worth of waste! Notice the different colour bags to denote the departments. The waste was kept in the factory all shift, so that people could see it mounting up.

- **Accurate** – based on good quality measurement
- **Timely** – the data must be as recent as possible
- **Information history** – there should be a history of data to demonstrate that the decision is based on a long-term issue and not a knee jerk reaction to a problem that is only short-lived.

If you want to manage something, first you have to measure it

This phrase occurred in the section that looked at KPIs and how day-to-day operations are managed. It is an old saying that without measurement, management is impossible. How can information be measured, on which to base our decisions?

Data capture

There will be an element of data that is recorded within the factory that may guide you. In the area of waste, because some of the seven forms are difficult to see, the data from the shop-floor may well be incomplete or missing some of the more difficult areas (see Figure 11.7).

Physical waste is easy to see and should be well recorded. However, the recording of physical waste is not always done diligently, especially if the department team is under pressure to perform. It has been seen in factories that waste is removed to the skips without being recorded, so that the performance looks better than it is. The other way that waste can be miss-recorded is the material that disappears down the drainage system of your food

Waste analysis – Packing department line 5.		
Total produced	15089	
Total waste	978	
Waste %	6.50%	
Reasons	Packs	Percent of waste
Faulty seal	321	32.80%
Light weight	124	12.70%
Faulty date code	78	8%
Fell on floor	204	20.90%
Other	251	25.70%
Total	978	

Figure 11.8　A typical waste analysis.

factory. The effluent outflow from your factory can contain large quantities of food materials that are unaccounted for if you simply weigh the black bags exiting a department.

Physical waste should be easy to spot, but can also be difficult. How about some of the most difficult of the seven wastes to identify?

War on waste

This could be from a specific study; you could tell someone to record the quantity and reasons for all the lost time on a machine. Or you could segregate the waste from a department to allow different causes to be captured. In the packing department, the following could be discovered (see Figure 11.8).

During the period of the study, of the 15,089 packs made, 978 were rejected, and this was 6.5%. At this point the factory manager was jumping up and down and demanding that the waste on Line 5 be reduced.

By capturing more information from a waste study, it can be seen that the big three causes of the waste were Faulty Seal (321 packs), Fell on Floor (204 packs) and Other (251 packs); there will always be an "other" category. You can see that with this extra information it was now possible for the team to tackle the main causes of the waste. Without the information that would not have been possible. In order to manage and improve something, first you have to measure it.

Sometimes it is easier to see the picture more clearly if the data is collected over a period of time. Today could have been an unusual day and seal faults are not normally such a big problem. What is needed is evidence of the real issues, and not opinions.

Another method of improving the picture is to put all the data into a diagram. Waste analysis is well suited to the use of graphs that track performance over a period of time (see Figures 11.9, 11.10 and 11.11).

Questions

- What is your data capture system for physical waste?
- Do you know the reason behind all of the major sources of physical waste in your area?
- Are you trying to reduce it with your team?

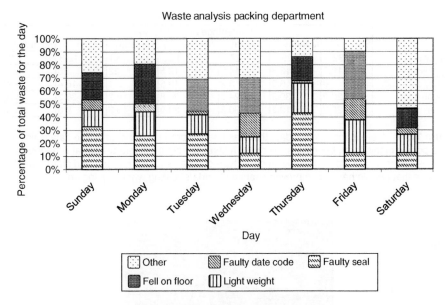

Figure 11.9   This chart shows the contribution that each waste cause made to the total waste on that day. It is easy to see from here that the waste picture is complex (it often is) but that the major issues in the week were faulty seals on Thursday, fell on floor on Friday and that, for some reason, "other" went through the roof on Saturday.

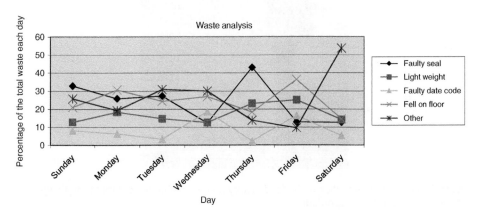

Figure 11.10   This is the same information presented on a line chart; this chart can tell us more about trends of the performance but the highlights are more difficult to spot.

What do we measure?

This is a problem in a complex food factory. If everything is measured, the business would sink under a huge wave of data (see Figure 11.12).

The quantity of information that could (and often is) generated in a food business is massive. To attempt to measure everything is not an option. Decisions need to be taken

Waste analysis

	Sunday	Monday	Tuesday	Wednesday	Thursday	Friday	Saturday
				Percentage of the total waste			
Faulty seal	32.8	25.8	27.2	12.3	43	12.8	12.6
Light weight	12.7	18.3	14.6	12.5	23	25	14.1
Faulty date code	8	6.2	3.2	18.5	2.1	16.4	5.2
Fell on floor	20.9	30.5	24.2	26.8	18.2	36.2	14.5
Other	25.7	19.2	30.8	29.9	13.7	9.6	53.6

Figure 11.11   Here is the original data that was used to produce the charts. The data is fine for some people in this form, but others will prefer a chart to help them interpret what is happening. The safe option, if you are communicating this type of data, is to use number and chart formats to ensure that people have an improved chance of understanding what is going on

Figure 11.12   This is the office of the food factory manager who thought that everything should be measured and recorded!

about what data is captured from the factory, in order to provide information for decisions and improvements.

When everything is important – nothing is important – anon.

The issue of too much information is a vital one for a business to tackle. The business needs to decide what information is required to maintain control and also improve the way it performs. Information adds no value to your products, and the collection and analysis of information simply adds cost.

Information and data is needed by the food factory for many reasons. The business needs to collect data for:

- **Food safety** – linked to the HACCP systems within the operation
- **Product quality** – linked to product and raw material specifications
- **Legal requirements** – there are some obvious areas where it is a legal obligation to collect and store information:
  ○ Health and safety
  ○ Employment law
  ○ Environmental
  ○ Financial
  ○ Average Weight controls for Trading Standards.
- **Customer service** – records are often kept in this vital area, such as shortages in delivery, late despatch. The numbers of products made falls into this area.
- **Cost control** – this is vital data to be collected, such as waste generated, number of people working on a production line, or giveaway on products that are too heavy when packed.

All of this data is needed and factories have systems to collect this data. Focus on the data requirements for controlling costs for a moment. As described earlier in this book, the costs of running a food business are made up of:

- Variable costs:
  ○ Labour
  ○ Raw materials
  ○ Packaging
- Semi-variable and fixed costs:
  ○ Management
  ○ Factory costs.

From the point of view of the food factory trying to improve and applying Lean Manufacturing techniques, there is a new category of costs that it is important to understand:

- **Actionable costs** – these are the costs that you can do something about. All costs are actionable, some are just more difficult than others. The issue of actionable costs also extends into the area of data collection and handling.

The issues of data collection are substantial, so data should only be collected if:

- It is useful:
  ○ The data is needed now or at a definite point in the future.
- It is actionable:
  ○ The data should be able to be acted upon to improve the business.

Other data should not be routinely collected, but can be the subject of special projects if required (see Figure 11.13).

Figure 11.13   Data is often collected manually in a food factory, is recorded on paperwork and is then collated at the end of each shift to ensure that all the required information is present. The paperwork is then filed away and will only see the light of day if there is an issue that requires investigation. "We have just had a consumer complaint about a foreign body in a product. Can you get the metal detection records for the 19th June afternoon shift?" In this case, the record may form part of the evidence for a defence in a court case, so it is important from that point of view.

Increasingly, as food factory machinery are linked to computer networks or handheld terminals are used to collect data, it is possible to have the records available without the cost of collection and storage (see Figures 11.14 and 11.15).

Useful and actionable

The types of data that might be needed in a food manufacturing business are:

- **Labour cost** – this can be divided into:
  - **Manning levels** – the actual number of people used in production compared with the number that should have been used
  - **Downtime** – lost time through breakdown of machinery, waiting for ingredients, changeovers between products
  - **Run rate** – the actual speed of the production compared with the standard speed
  - **Reject level** – quantity of product rejected, each rejected pack has had some labour added to it as it is made. By rejecting the pack that labour cost is wasted
  - **Absenteeism** – people who did not show up for work. This can have a significant impact on performance in a factory

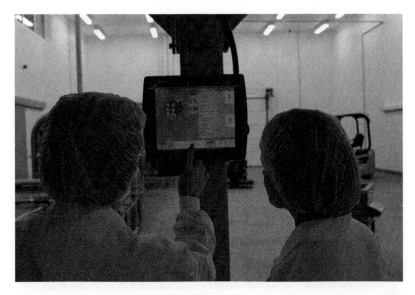

Figure 11.14   Handheld terminals or machine control screens are linked to ERP systems used for data collection. The data is uploaded to the company computer system, so the data is immediately available. If an hourly check on an HACCP in the factory is carried out ten minutes late, the business knows about it. If there has been a rise in the number of rejects in packing because of faulty coding, the business knows about that too. That last changeover was performed three minutes quicker than schedule, the business knows about that too. No paperwork, the factory captures data on KPIs and other factors, measures and records performance and completes control records for traceability and legal requirements. All with a minimum of manual inputting and recording.

○ **Productivity** – the actual packs per man hour compared to what it should be
○ **Pay rate per hour for employees** – employees cost more than their wages. National Insurance, holiday pay, sick pay, subsidised canteen and cost of work wear and protective clothing all add to the hourly rate that a company pays for its staff.

Other areas of vital data collection allowing a level of control over the costs of a food manufacturing business are:

• Raw food material and packaging material cost:
  ○ **Waste** – good material that does not make it into the products, but ends up down a drain, in a waste skip or in a recycling point
  ○ **Yield** – where materials are converted in the process, there will be a yield effect. Put 200 kg of chicken meat through a cooking process and its weight will reduce to 168 kg because of the cooking losses. Hot sauces from a cooking vessel will lose weight through evaporation during cooling.
  ○ **Average pack weight** – if a pack is supposed to be an average of 200 g and at the end of the shift the average is 206 g, then on average 6 g has been given away to your customer in every pack. Your customer did not expect this extra product and neither did they pay for it. It has cost extra to make these packs.

Figure 11.15   Handheld terminals or machine control screens are linked to ERP systems used for data collection. The data is uploaded to the company computer system, so the data is immediately available. If an hourly check on an HACCP in the factory is carried out ten minutes late, the business knows about it. If there has been a rise in the number of rejects in packing because of faulty coding, the business knows about that too. That last changeover was performed three minutes quicker than schedule, the business knows about that too. No paperwork, the factory captures data on KPIs and other factors, measures and records performance and completes control records for traceability and legal requirements. All with a minimum of manual inputting and recording.

○ **Reject level** – it is obvious that the waste of raw materials and packaging has a cost associated with it. But remember that if the material has been partly or fully processed, then the value of that material has increased above the purchase price. Also there is a huge and increasing cost associated with waste disposal, including handling and transport of waste, as well as skip charges and landfill taxes.

Back to the original question: "What should be measured?"

Even by restricting the measurements to those that are useful and actionable, there would still be too much data collected from a business in most cases. To reduce the data collection still further, it is possible to look at:

• **KPIs** – these are measurements that are important in their own right, but also give strong indications of any issues that may need to be addressed. The important thing is to use the data from KPIs to manage the business, without letting the business sink under a wave of data that is neither useful nor actionable.

Data into information

Looking at a page of data collected from the shop-floor, it will not be easy to see patterns and trends in the numbers. It is important that data is converted into information, so that it can be easily understood and interpreted (see Figure 11.16).

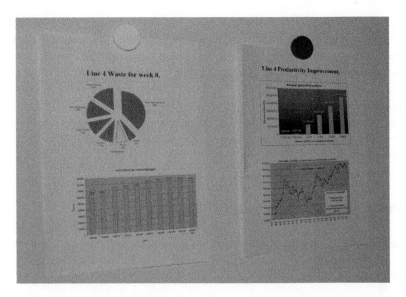

Figure 11.16 Other, sensitive information was removed from this board before this picture was taken. The use of graphs to convey information is easily understood in a few seconds and can then be acted on to improve performance. Trends in performance can be shown as well as recent results.

How do we manage the data in a business?

The first rule is "information not data". It is vital that data is processed in such a way that it is converted into useful information to make decisions and control the operation. In order to convert data into information, there are a few simple rules that should be applied in a Lean Manufacturing operation.

A measurement means nothing unless it is set against an expected figure or target. All data collection should take place with a target being in place. That way it will become obvious if the data is drifting from the target. SICs are always set using this as a base:

- Data, in isolation, is just a list of numbers. One of the principles of the 5S system is the visual factory. If the data collected is made more visual, it will be easier to interpret and as a result, corrective action will be quicker to help keep the process under control.
- Data can be turned into information by:
  - **Drawing a graph with a target line** – this can be done manually but more often now as data is collected by computers that monitor the process the display is in the form of a graph with target lines. If it works for data that is collected automatically, how about the data that is collected manually? That too should be displayed visually if possible.
  - **Calculating an average, range or Standard Deviation** – this is a useful way of reducing the quantity of numbers that can be collected during a shift. If an average and range is calculated for the data collected, this will allow easy comparisons. "Last week we averaged 24 minutes per shift in lost time, this week so far its 38 minutes per shift on average – what is going on?"

○ **Comparison in trend analysis with historical performance** – if data is collected over several weeks or months, it is possible to display the information in the form of a trend graph. It can be seen if performance is improving over the long term and again this can aid decisions for the business, its managers and teams. "The trend over the last 12 months is showing that the number of breakdowns is reducing but that the average lost time has increased slightly – what is going on here?" "The trend in our output per man hour is showing a good improvement, except on days where the order drops unexpectedly – we should look at ways of improving our ability to respond to a drop in orders."

Key Performance Indicators KPIs

KPIs are vital in the collection of information in spotting the seven wastes. They allow for efficient data collection of vital measures in the business, but do not cause a tidal wave of data that would occur if everything were measured.

How do we decide what is a good KPI for our business, department or shift?

Deciding on the KPIs is a logical process. Various tools can be used to identify the KPIs for your business. Among these are:

Pareto analysis

- Pareto analysis is sometimes called the 80:20 rule. For example:
  ○ 80% of your costs come from 20% of your ingredients
  ○ 80% of your absenteeism comes from 20% of your employees
  ○ 80% of the rejects come from 20% of the possible reasons.

  Products may suffer from different defects but, for example:

- The rejects from a packing department are caused by a number of reasons:
  ○ Open seals
  ○ Faulty code
  ○ Damaged pack
  ○ Fallen on floor
  ○ Wrong product in bag
  ○ Underweight
  ○ Overweight, etc.
- 80% of the rejects will come from a relatively low number of the possible faults, for example, most rejections are due to seal problems, next most common cause is underweight packs, etc. If the business collects data on the top three or four reasons and acts on the results, it will be tackling 80% of the waste problem. Pareto analysis can be applied to almost any area where data is collected (see Figures 11.17 and 11.18).

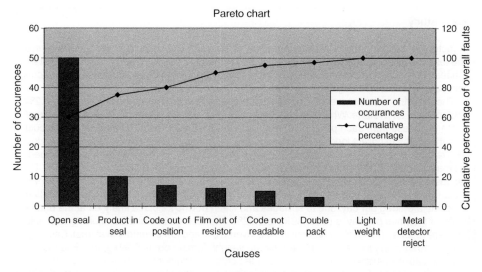

Figure 11.17 It can be seen from this chart that the top three or four reasons for waste account for almost all waste. The bottom three or four contribute almost no waste. If you are trying to reduce waste, your time should be spent on the top three or four reasons. Working this way will give the maximum effect for your efforts.

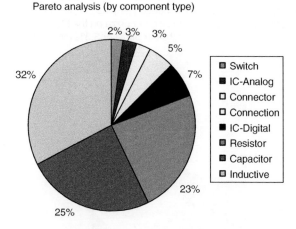

Figure 11.18 Pie charts can also be used to make it more obvious where you should focus the improvement activity. The big wedges here are likely to give a big "bang for your buck". If you look closely, this is a Pareto analysis chart from a micro-electronics factory, just to show that everyone is doing this type of stuff to make their business better. This was one was the causes of failure of a printed circuit board. You can see here that there is a top three causes of failure; they should be improving first.

Having done some Pareto analysis, it is then possible to design a KPI and data capture system to ensure that you have the information available to make decisions and monitor the effects of those decisions.

Having completed the Pareto analysis, it is easy to see where the control is needed in the factory. Making giveaway a KPI and then challenging the team to improve the number

Case study

A factory is using too much cheese in the manufacture of its pizzas, but has no idea where the loss is occurring. The team decide to collect some data to see if the loss can be spotted. Cheese is watched at every step and all cheese handling is monitored for a few days to collect data for a Pareto analysis of the loss.

The following was discovered. The cheese that was delivered from the supplier was 0.2% underweight on average. When the cheese was on site, the QA department took a sample of 200 g of cheese from each pallet of 1000 kg, so that is a 0.02% loss. The staff canteen would visit the stores to collect food for the meals it was preparing and they would take 2 kg per day. The factory uses 5 tonnes per day, so that is a 0.04% loss. When the cheese was unwrapped, a small quantity of cheese stuck to the packaging, say 20 g per 20 kg block, which is 0.1%. So far, losses of 0.36% have been spotted and the cheese has only just come out of the wrapper! Next the cheese is grated into tubs of 10 kg per tub. The total losses on the cheese-grating machine are 36 kg per day, due to spillages and clean downs for blade changes at the end of shifts. If 5 tonnes per day is used, 36 kg is 0.72%. The tubs go into a chiller store to wait until they are required (this factory has not yet discovered JIT and low inventory systems). Each tub weighs 10 kg going into store but the average weight coming out is 9.99 kg, that is a loss of 0.1% because of evaporation of moisture in the cold environment. The cheese gets taken to the pizza assembly line when required and is tipped into the cheese depositing system. The tipping operation causes some spillage and the total at the end of the shift is 42 kg, that is 0.84%. The empty tubs are not quite empty, as the grated cheese sticks to them and they are sent to tray wash with a total of 22 kg in the 500 tubs of cheese used per day, which is 0.44%. That is only 44 g per tub on average. Now the cheese has to be put on the pizzas, even after 2.46% of it has already been lost. The cheese depositing system is a bit erratic, so to make sure that the pizzas have enough cheese, the machine is set to put 2 g extra per pizza to prevent lightweights. The required weight of cheese per pizza is 50 g but an average of 52 g is placed, to get somewhere between 50 and 54 g per pizza on the 83,000 pizzas made from the 5 tonnes of cheese. That is a giveaway of 2 g per pizza or 3.3% (see Figure 11.19).

Figure 11.19   This is our cheese depositing machine that is not performing well enough to be controlled tightly. Some improvement needed here (picture courtesy of Ishida Europe).

Some good work by the improvement team left no stone unturned in looking for the lost cheese. Ten causes of the losses have been identified and this is an ideal time to break out a Pareto analysis chart of some description to make the data turn into something that everyone can understand. With the examples given here, you can see how some of the losses are hidden and only appear when an investigation looks at the detail of what is going on. Of all waste cheese generated, the only areas that are generating black bags with cheese in are the grating operation and the tipping operation. All the rest of the waste leaves the factory in other ways (see Figures 11.20 and 11.21).

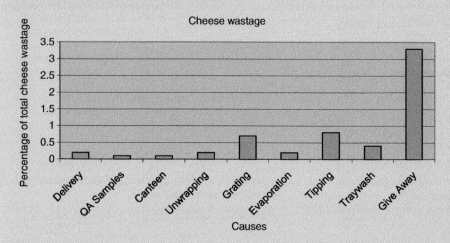

Figure 11.20   Here is a chart of the cheese wastage where you can see the big offender straight away. A Pareto analysis chart is usually organised in descending order, so that the biggest causes are to the left and the least significant are to the right.

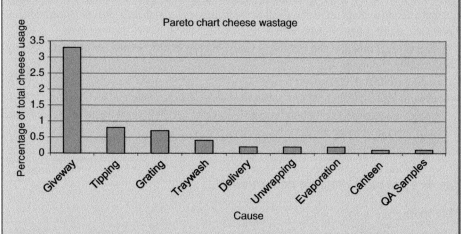

Figure 11.21   The causes are now in descending order, so it is easy to pick off the top three or four reasons to tackle. In this case the giveaway, the losses on tipping cheese into the machine, losses on the grating operation and the quantity of cheese left in the tub when it is "empty". The rest of the causes are so small that even if you could solve them, it would still leave a large problem. They should be addressed after the team has had a go at the first few with their improvement ideas. Though a quick call to the supplier would be simple to do and should give a quick result!

would be a great idea. But how about measuring the cheese waste on grating and monitoring it at the start of shift meeting. Get a team to look at methods of reducing cheese waste here. One area of interest in this case study is the impact of rushed or sloppy work on the tipping job to get the cheese into the depositing machine. First, cheese misses its target and ends up on the floor and second, there is a substantial quantity of cheese left in the tubs that also contributes to the waste significantly. Once an issue in the factory starts to be investigated and starts to generate many causes, a bit of Pareto analysis will help you focus on the top few causes that will account for 80% of the issue. The 80:20 rule is alive and well in your factory.

Questions

Think now of three or four areas where you might try to use Pareto analysis:

- Breakdowns
- Absenteeism
- Product rejects
- Ingredient and packaging usages.

It is not until you try this powerful technique that you will see how it can focus you and your team on the causes that are really holding back your performance.

Cause and effect charts

- The next method for deciding on appropriate KPIs is a cause and effect diagram, sometimes called a fishbone diagram. It shows all the links between variables in the operation and may point to a useful and actionable KPI to help manage and improve the operation.
- Fishbone diagrams have no statistical basis, but are excellent aids for problem solving and trouble-shooting.
- Cause-and-effect diagrams can:
  - reveal important relationships among variables and possible causes
  - provide additional insight into process behaviour and help identify the important data to collect (see Figure 11.22).

How is a fishbone diagram constructed?

Basic steps:

1 Draw the fishbone diagram.
2 List the problem/issue to be studied in the "head of the fish".
3 Label each "bone" of the "fish". The major categories typically utilised are:

- The 4 Ms:
  - Methods
  - Machines

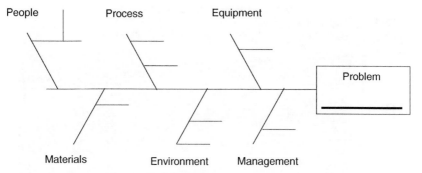

People   Process   Equipment

Problem

Materials   Environment   Management

Figure 11.22 One example of a fishbone diagram that can be used to help identify the causes of a problem. The diagram is usually completed by a team of people working together, who bounce ideas around and come up with a shared opinion of the things that contribute to an issue. It is normal in the food manufacturing industry to consider the six headings above, but in specialist areas others may be appropriate.

- ○ Materials
- ○ Manpower
- The 4 Ps:
  - ○ Place
  - ○ Procedure
  - ○ People
  - ○ Policies
- The 4 Ss:
  - ○ Surroundings
  - ○ Suppliers
  - ○ Systems
  - ○ Skills

Note: You may use one of the four categories suggested, and then combine them in any fashion or make up your own. The categories are to help you organise your ideas. You will notice that the diagram above uses six categories to try and capture the possible issues around a problem.

4 Use an idea-generating technique (e.g. brainstorming) to identify the factors within each category that may be affecting the problem/issue and/or effect studied. The team should ask, "What are the machine issues affecting/causing…"

5 Repeat this procedure with each factor under the category to produce sub-factors. Continue asking, "Why is this happening?" and put additional segments into each factor and subsequently under each sub-factor.

6 Continue until you no longer get useful information as you ask, "Why is that happening?"

7 Analyse the results of the fishbone after team members agree that an adequate amount of detail has been provided under each major category. Do this by looking for those items that appear in more than one category. These become the "most likely causes".

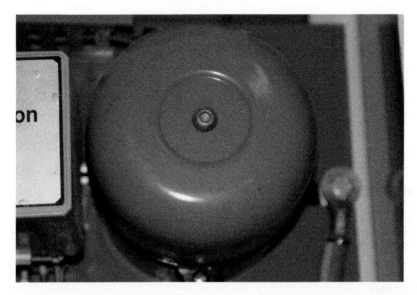

Figure 11.23   The alarm bells that will ring if the SIC system shows that the customer is going to be let down unless something is done quickly in the factory. Someone will be monitoring the main causes of a poor service record, so if it starts to happen again, decisions can be made and corrective action can be taken.

8   For those items identified as the "most likely causes", the team should reach a consensus on listing those items in priority order, with the first item being the "most probable" cause.

For example, supposed there is an issue in the factory with a low level of on time/in full delivery to our major customer. The situation is a complex one and is partly due to under-performance in the factory and partly due to wrong data input into the planning computer system. A fishbone diagram could help the team sort through all the issues and make sure that all the angles are covered. At the same time, it cuts through the normal arguments that can occur when a team is under pressure. The production team blame the planners for not planning enough production, the planners blame the QA team for rejecting the stock that was left over from yesterday, and the QA team blame production over the lack of control in the packing area.

This finger pointing has to stop and someone has to come up with a solution or the customer will go elsewhere for their product. Cause and effect analysis or a fishbone diagram can get people to talk around the issues and identify the main causes of the poor service level. KPIs can be installed to monitor those main causes and they can form part of a SIC system to make sure that alarm bells start ringing early if the KPI is drifting off track (see Figure 11.23).

Criticality

The final method for the determination of appropriate KPIs is how critical it is to the business?

- How critical is the control and data to the costs of the business. For example, the following may be examples of KPIs in a warehouse that allows the manager to control the business:
  - **The "stock losses" of the top ten expensive raw materials** – stock losses can be caused by wastage in the factory, poor control of stock rotation resulting in raw material disposals from the store as they are too old to use, and planning issues where too much of a material has been purchased and as a result is not used before it runs out of life and is no longer fit for use. The final area in a warehouse comes under the heading of unrecorded withdrawal from the store. Material is taken to be used in the factory without the knowledge of the stores or the planning team. This can happen to cover a problem in the factory where waste has been generated because of a process problem or it could be withdrawn by theft and it is about to disappear off site. The control of stock in the warehouse is critical, as the material is a high proportion of the overall cost of the food made and a lack of control over the stock can result in disruption in the factory, where costs can soon go out of control because of unplanned stoppages.
  - **Packaging usage rates as the business approaches a design change or delisting** – it is critical that packaging is available for the products being manufactured. That is an obvious statement; you cannot sell food in an unpackaged state, with the possible exception of crusty bread in a bakers shop, but even then they put it in a bag! The packaging contains so much information about the product that it is often specific to a particular product. Most of the time the availability of packaging is not an issue in a food factory, because it has a long life and stocks can be held to ensure continuity of supply. That said, a few companies have switched to JIT delivery but the packaging stock is usually held at the supplier rather than the food factory. There are periods where it makes sense to reduce the packaging stocks to as low as possible.

    When a package is replaced or a product is de-listed or switched off of the promotional pack offering "World Cup" T-shirts, and will not be manufactured again, it is a good policy to run the stock down to a very low level to minimise the value of the stock just before the change.

    In order to have a "soft landing" and ensure the stock and usage are managed closely, it may be worth making the stocks of these items a KPI for the period up to the change.
  - **Stock levels on long lead-time materials** – some materials used in the manufacture of foods can take a long time between placing an order and receiving delivery. These long lead time materials may also be on a list of KPIs to ensure sufficient stock is either in the warehouse or on order to ensure continuity of supply to your customer (see Figure 11.24).
  - **Productivity during a period of promotion** – productivity will most likely be a KPI in most food businesses for the whole of the time, as labour has such a large contribution to the overall costs of the product. During periods of high activity, productivity measures can be increased to give more detail of the performance in this area. During a large promotion, it is possible that the factory is under a lot of pressure from a capacity point of view. The need to keep a close eye on the productivity performance

Figure 11.24   Long lead times can force you into ordering more stock when you have yet to get the last order delivered. Planning and control of critical stock items can justify them being monitored by a KPI.

Figure 11.25   During a BOGOF promotion, it may be worth measuring productivity for each production line rather than the whole department.

means that more measurements are typically taken. Rather than measuring productivity at the end of a shift, it can be measured each hour. Rather than measuring productivity of a whole department, it can be broken down into individual production lines (see Figure 11.25).

○ **Downtime during a period of service issues** – this is an obvious one. Productivity would be measured most of the time as a KPI and maybe an OEE too, but when the pressure is on and machine breakdowns cannot be afforded, it would be a great idea to focus on keeping the machines running by making the downtime a KPI. Once the downtime is monitored in this way, it will be much higher in the minds of the team and they will work together to make the KPI as good as possible.

KPI impact

KPIs can have a massive effect on the performance of a business, but also on the way that performance is created. KPIs help in the close management of the detail of a factory, not by monitoring everything (that would be too costly), but by monitoring carefully selected indicators of performance:

- The KPIs for a business indicate to everyone what are the important parts of a business that are monitored consistently. This includes the factory staff, the support staff, the senior management and also the customers of the business.
- The KPIs may not show the complete picture, but are the "radar screen" for the management of the business and clearly show by comparison with targets if the business is under control.

The careful monitoring of KPIs will give warnings of problems and opportunities emerging on the horizon. Waste percentage has been creeping up on line 2, so the causes must be identified. Labour cost per tonne is really good at the plant in France, so there is a need to look at what they are doing and try and repeat it here. Changeover times are a bit erratic, so the methods need to be examined and some retraining be done if necessary:

- The impact of the information contained in a KPI is magnified if all people in the business have access to the results and they are presented in a usable way. Lean Manufacturing is about taking thousands of tiny steps to make continuous improvements. The more people that are involved in that process, the faster the improvement will be.

Daily/Weekly Operating Report (DWOR)

- A DWOR is one method of ensuring that KPI information is captured and used to inform the business. It becomes immediately apparent if a KPI is missing its target. Shift by shift and day by day, the KPI information is brought together and recorded. Each performance number can be immediately compared with the ones around it. Are they better than the target? Are they better than yesterday? Are they better than the other shift? The DWOR shows immediately if a performance is off target or that there is a trend in the results.
- KPIs for each day are built up into a total for a week and can easily be converted into graphs and trends in performance. It can become the focus of daily production meetings and ensures that all occurrences of poor or good performance is recognised (see Figure 11.26).
- Daily performance for each KPI is instantly compared with the previous day, as well as the target for each of the KPIs. Actions can be created to tackle any difference. The use of a DWOR gives improvement teams immediate feedback, so they can see if their work is having the desired effect. This will inform the team that they are moving in the right direction and encourage them to go further.

Factory ABC – Daily Weekly Operating Report

Week Comm
Shift Manager

KPI	Target	Sunday	Monday	Tuesday	Wednesday	Thursday	Friday	Saturday	Total
OEE									
Labour Efficiency									
Waste									
Quality									
Accidents									
Absenteeism									
Energy Consumption per tonne									

Figure 11.26 An extract of a DWOR for the ABC Factory team. Data entered onto the DWOR can be quickly compared to target. Running totals and averages can be calculated for the week so far, to ensure that focus is quickly placed on the areas that need attention to prevent a bad day becoming a bad week.

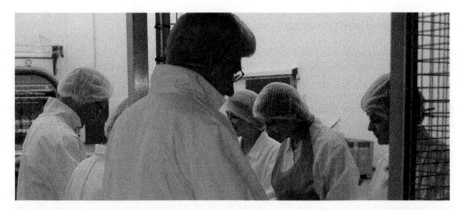

Figure 11.27   The departmental manager in this factory needs to ensure that the KPI system is fully working. The data entered is correct and the team is taking action when the performance falls below target.

- Over a period of a few weeks, a history of the performance can be built up and trend graphs can be created so that the KPI can be managed and controlled. In the food industry it is often difficult to remember how things were before. Everyone is so focused on the future that it can be useful to have performance records that look back, so that people can get a feeling for the progress made in the last month or last year.
- Targets can be set for the KPIs and actions and investigations can result from performances that fail to meet the required level. Ensuring that all underperformance is acted upon, rather than simply be accepted, is one of the main functions of a KPI system. Without the corrective action, nothing will change. If you do the same thing tomorrow as you have done today, why should the result tomorrow be any different? Just saying "Today was a bad day – tomorrow will be different" is not sufficient. Managers and teams have to make sure that tomorrow is different, either by preventing the mistakes that occurred today or by improving the performance in some way (see Figure 11.27).
- Gaps in the data are immediately apparent and the quest for reliable data can be made consistent across all shifts and departments. Missing data is one way in which a KPI system can fail. The creating of the information needed to calculate a KPI can be a large task. The collection of KPI data must be made simple and routine to ensure it is available when required and is complete and reliable. The automatic collection of some of the information is one way of reducing this burden on the team but if the team is well motivated and sees the importance of the KPI system, it will be a task that will yield great benefit to the factory.

That concludes the section on collecting information from the factory and ways of using that data to point towards opportunities to improve. Our thinking has been applied here to the seven wastes, but a lot of what has been already been done applies to other ways of managing a food production operation.

Comparisons of performance

Benchmarking is a technique that allows comparisons to be made inside and outside of a business, to discover "what good looks like". A factory may believe it has the best level of performance in the country, but it cannot be sure until it compares itself with others:

- Once a good system of measurement is set up in a business, it is possible to gain even more information by a process called benchmarking. KPIs are compared and differences are investigated. A factory running with very low wastage may think it is the best at waste control, until it benchmarks with a factory that has waste levels that are even lower. The same applies from production line to production line, department to department, or even machine to machine.
- If measurement methods are consistent across several businesses, shifts or departments, then benchmarking can occur. The secret to good benchmarking is to ensure that the comparisons are valid and you are not comparing apples and bananas. If waste is differently defined in two factories, then benchmarking that KPI may be difficult. Maybe one factory defines everything that is not sold to the customer as waste. Stock sent to the company shop or samples taken for taste panels are all waste. Another company may define waste as all products that end up in landfill or recycling skips. Because of the great advantages that are shown by benchmarking, there have been some KPIs that are defined in a way that makes them universal to all businesses. How OEE is designed to be a KPI for all manufacturing businesses, and how it can be used to benchmark across widely differing factories and industries, will be gone into more detail later.
- This is where data is shared to improve performance, by letting the business know what is possible and what is achieved elsewhere. Looking at your own performance and comparing it with your target and your previous record is all well and good, but it is important to look around sometimes and see how far the KPI is pushed elsewhere.
- World Class Manufacturing is the term given to the performance of the top 5% of manufacturing companies in the world. These companies make everything from motor cars to micro-electronics, from prefabricated houses to jet engines. Some World Class companies make food and drink products, so there are some very high performing companies in our sector.
- Each year there are benchmarking competitions organised by various trade bodies and universities, to try and give factories an independent view of their performance. Most of these competitions ask a bank of questions and then produce a league table type report. In order to encourage companies to share their information, the reports are confidential. You know that your biscuit factory is fifth in the table for the quantity of work in progress as a proportion of the tonnes per week output in biscuit factories. You do not know the name of the other companies, but you do know what good looks like in your part of the industry. Targets can be set and plans made to get you to the top of the table (see Figure 11.28).
- Shared information within a group of factories is equally as valid for improvements to be made. In this way, an innovative improvement in one factory will be detected by the sister factories as soon as it hits the shared KPI data. The impact of the work of an improvement team can be magnified as other departments and factories get to hear about it.

All benchmarking exercises require that the measures are identical in the businesses being compared. This can sometimes be difficult to achieve, but there are real benefits if a reliable system can be found.

Figure 11.28   The performance of this factory can be compared to others to try and see what good looks like. Benchmarking competitions are a fun way to gain an independent view, where you know that comparisons are made between apples and apples, and not apples and bananas.

Benchmarking can allow a quantum leap in thinking and give a business a lot of information and motivation to improve performance:

- How good are we?
- How fast are they?
- How many engineers do they have?
- How quickly do they changeover?
- What is their wastage?
- How many customer complaints do they get?

Questions

- How does the performance of your factory compare to others trying to carry out similar work?
- Does your business use benchmarking as a technique to know where improvements could be made?

Overall Equipment Effectiveness (OEE)

One common measure used by companies to benchmark themselves is the OEE. It is a good KPI that allows different industries to compare their performance. OEE measures of the performance of a machine or production line and is not concerned with labour cost, so can be used to compare machine performance in a wide variety of situations. In simple terms, the OEE is the quantity of good product made by a machine or production line compared with the quantity that theoretically could have been made in the same time.

OEE is one of the most widely used measures of manufacturing performance. An OEE of 85% is thought to be World Class manufacturing and means that the machine or production line made 85% of the theoretical maximum it could have made. The total of the losses is 15% and accounts for all machine stoppages (Availability Losses), all the times the machine was ready and running but no product was made (Performance Losses) and all the times a product was made but was subsequently rejected (Quality Losses).

Case study

A pizza production line was able to run at a speed of 60 pizzas per minute. The shift is 480 minutes long (8 hours), so in theory the line should make 28,800 pizzas in that time. During the shift there were some availability losses; the line did not start until 8 minutes into the shift because of an electrical fault. The line changed over 3 times onto different varieties of pizza that lost a further 32 minutes. Our 480 minute shift included 40 minutes of availability losses (8.3%). During the shift, the line slowed down because of difficulties with one of the ingredients. It ran at only 40 pizzas per minute for a period of 3 hours. That meant that only 7200 pizzas were made during that 3 hours instead of 10,800, that is a loss of 3600 pizzas or the equivalent of 1 hour of production. There was a performance loss of 1 hour during the 8 hour shift (12.5%). Finally, during the shift, 3200 pizzas were rejected. The time taken to make those 3200 pizzas was 53 minutes, so that is a quality loss of 11.1%. The total losses for that shift were Availability 8.3%, Performance 12.5% and Quality 11.1% – Total is 31.9% giving an OEE of 68.1% (see Figure 11.29).

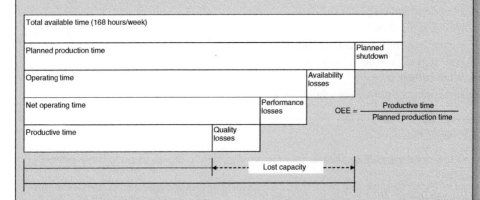

Figure 11.29 OEE is a great measure of performance, as there is no hiding place from the calculation. No excuses can be made. If your OEE is below the target you have set yourself, the loss of performance will be contained within availability, performance or quality losses. There simply is nowhere else it can be. Improvements to the OEE of a machine or production line are very good news for a factory. Getting your output closer and closer to the theoretical maximum is the very heart of continuous improvement and Lean Manufacturing.

Some hints on opportunity spotting

Business improvement methods are only useful to a business if the opportunity for an improvement is spotted and the business accepts the need to change. The first step is to spot the opportunity. There are well established methods in spotting opportunities that will act as the toolbox of a manager who wants to improve the way that a manufacturing operation works.

Performance comparisons

If a business measures its performance in a set way, it can use the results to spot an opportunity to improve:

- KPIs are often used in comparisons
- Comparisons can be made between the current level of performance, and:
  - Last week, last year
  - Performance of another line in the factory
  - Performance of another factory
  - The calculated maximum performance
  - The design performance of the line.

There are many performance targets that can be used as your "yardstick" or benchmark. The key here is to make sure you keep your eye on the same target for a reasonable length of time. A moving target is very difficult to hit.

Once a difference has been spotted, comparisons of KPIs can be used to identify the cause of the issue. By asking questions such as "What happened five weeks ago? The waste has been increasing since then," to discover that five weeks ago saw the launch of a new product or the installation of a new machine. "Why does our sister factory have less performance losses?" to discover that the sister factory has been using a new type of display screen in their production area, which gives the speed of the line information in numbers that are 300 mm high!

Double handling

Double handling is where a product or material is touched more than once while it is in use. The touch could be by a person but could also be by a forklift truck.

Three of the seven wastes are unnecessary motion, waiting time and excessive transport:

- All of these are included in the double handling of a product.
- Ideally a "one touch" system should be operated to ensure that product flows through a process and is not constantly put into storage locations for short periods, only to be handled again.
- Remember, if you see a product that is not moving and not being worked on, then double handling has occurred. You have paid someone to put that product in that position and you will pay someone else later to retrieve it.
- It is very rare that a factory can work without some level of double handling; as with all things in Lean Manufacturing, the aim is to work towards perfection in a series of steps that will allow you to continuously improve your process and performance.

High waste and rework levels

Improvement opportunity spotting method 3 is to look out for areas of your factory that are generating high levels of waste or reworking. If a business is suffering high wastage levels, there will be an opportunity to improve:

- If a business is suffering from not getting it right or having to rework materials due to production issues and errors, then an opportunity is waiting.
- Management by "looking in waste bins" is not glamorous, but will help narrow a large waste stream into a few opportunities to reduce cost.
- Remember that this waste comes in seven different forms and is not just the stuff in the bins and black plastic bags. Waste could be production staff that are ready to work but there are no materials available. Waste could be a machine running at too slow a speed. Waste could even be people who are late back from their tea break.

Customer service failures

- When a business lets down its customer with poor service levels or quality rejections, it is an indication that improvements in financial performance could also be achieved. The costs of a service failure are many and varied. There is the direct cost associated with the loss of a sale to your customer, but this is sometimes made greater by the customer making an additional charge on your business for the loss of profit at the retailer. The biggest hidden cost is one that appears when your customer takes their business away from you and places it with a competitor because of service level issues and your performance in this vital area. Of course, the shoe can easily be on the other foot and your business could gain additional sales because of service level issues at a competitor.
- Root Cause Analysis is needed to ensure that the improvement opportunity is not missed. This is always a good time to get the team to think about the issues leading to the service failure, using a fishbone diagram to make sure the main causes are identified and sorted out.

High accident rate

Everyone coming to work has an expectation that they will be safe. Your company has a legal obligation to protect the health and safety of everyone in your factory. An accident or near miss within your business needs to be acted upon to ensure that everything reasonable is done to prevent a reoccurrence:

- A higher than normal accident rate is an indication that controls in the factory are not where they need to be. Accident rate is a useful indicator of the way that manufacturing operations are organised and controlled.
- High accident rates will have an impact on staff performance as well. Motivation and morale will be greatly impacted by a workplace that is seen as hazardous by the people in it, who need to feel safe if they are to work at their best. Asking someone to think about the quality and efficiency of their operation will not work if that person feels unsafe. A low accident and near miss rate is an important first step in getting the manufacturing teams to be creative and come forward with improvement suggestions.

Figure 11.30 This automated meat packing line should have few issues with strain injuries but automation brings its own safety risks and these obviously have to be managed correctly (picture courtesy of Ishida Europe).

- Lost time accidents will also distract the management away from improving the business. Where an accident results in someone unable to work for a period of time, it becomes a major task for the management and may stop them thinking about performance improvements. Too many lost time accidents can be a major cause of the pace of improvements slowing.
- Strain injuries also indicate that the workplace is in need of improvement from an ergonomic point of view. If the ergonomics of a workstation can be improved, the operative will be able to achieve more with less effort, productivity will improve and that can be very important in some labour intensive processes (see Figure 11.30).

Stock levels

When still looking for opportunities to improve, it should be noticed that there are large stocks in the business. Stock of raw materials, packaging, work in progress and finished products should all be minimised in a factory applying Lean Manufacturing techniques:

- High stocks of anything in a business are again one of the seven wastes and indicate a lack of the required controls. Low stocks can also lead to waiting time and availability losses. JTL is costly, JIT is what is required.
- High performing businesses are run with minimal stock levels in raw materials, packaging, work in progress and final goods.
- If you spot that stock levels are not correct, then there is an improvement opportunity; by reducing the stock you are freeing up space in the factory as well as freeing up cash for your business.
- Stock that is held out of its normal location is also an indication that stock control could improve.

The secret of running a food business with the minimum size of stocks possible is to plan the stocks that are required at each step and then set the factory up to replenish those stocks only when required. A system of stock management called Kanban is commonly used to help in this area of factory and process design. Traditionally a factory would be designed to push raw materials through a process and the final product would pop up in the despatch warehouse. A process would be started and would keep going, almost no matter what was happening at the next stage of the process.

Case study

For example, in a biscuit factory, the dough was be mixed and the biscuits were formed and baked and it was up to the packing section to cope with what was coming at them, no matter what the performance of the wrapping machines. On a good day the factory would work well but on a day where the wrapping machines were playing up, there would be an accumulation of biscuits that had been baked and not packed. These biscuits were packed into tubs so that the wrapping machine could be repaired. For example, a stock of unwrapped biscuits suddenly appeared in the middle of the factory. All of those biscuits had to be double handled into the tubs and then back out again. The labour cost of packing those biscuits into the tubs was not planned. The storage location for those biscuits was not planned. These are the consequences of a Push factory.

Kanban systems try to address some of the waste issues that are inherent in a Push system of operation. The trigger to carry out work is given by the machine or department, signalling that they are ready to receive material. Only when the receiver indicates that work is required is that work carried out. As a result, the factory starts to operate a Pull system, where products flow through the factory more smoothly with less waste in terms of double handling and unplanned storage.

There are many ways to coordinate this Pull system of the operation and one of them is called Kanban stocks. In a Kanban system, small planned levels of stock are held before each section of the process. When the Kanban stock reduces to a trigger point, a signal is sent for the stock to be replenished. It does need a great deal of planning and thought for an effective Kanban system to be established. As with all things in Lean Manufacturing, once a system like this is installed, it can be gradually improved by the team and the impact of the system will be able to be tracked in the KPI performance of the area.

Moving from Push to Pull systems of operation in a factory can be a real challenge, as it is a totally different way of operating but if it can be achieved then there are great business benefits (see Figures 11.31 and 11.32).

Apply the lesson of the beer drinking to the factory and a two-bin system could be used to control the supply of sauce to a depositing machine. The guy mixing the sauce used to work at his own pace and would often get in front of the required rate. The excess stock he produced would be a source of waste because it had to be stored somewhere, it had to be double handled, it used up sauce containers, etc. Sauce mixing was pushing sauce to the production line. Now a two-bin Kanban is introduced. The signal for sauce to be made is the presence of an empty container next to the line. The sauce is then mixed and delivered straight to the line JIT, before the line runs out. Perfect Pull system. No stock, no double handling, no risk that the stock would not be required.

Figure 11.31 In this FIFO store for a small food factory, each of these pallet racks holds a lorry load of packaging (26 pallets). The Kanban signal to order more materials is when there are three empty racks. The order is placed for 9 lorry loads and the lead time for the delivery means that the factory is down to its last 26 pallets when the new consignment arrives. JIT Kanban stocks can be on a large- or small-scale.

Figure 11.32 You have taken the team to the pub to celebrate the success of the department productivity improvements over the last month. Here is an example of a two-bin Kanban system that each of the team has on the table in front of them (these people are very good at Lean Manufacturing techniques). The queues at the bar are bad and it is taking 15 minutes to get served. Coincidently, you are drinking at a rate of four pints per hour. The signal to order a new pint is one empty glass in the two-bin Kanban you have set up. Your new pint arrives JIT and you continue drinking at your desired rate. Really efficient (well for the first two hours anyway)!

Wrong line balance

Just watching a team of people working in a manufacturing operation can allow you to spot an opportunity to improve:

- Where areas of a production line appear to be working too hard, while others are just ticking over. This could be due to the different skill levels of the people carrying out the task or, in an automated line, the different production capacities of the equipment. But is could also be due to the line being out of balance. It will never run smoothly, because of the work it is being asked to do. There is room for improvement here, where a line appears to be struggling in some areas but not others; it is worthy of investigation by the team to see if there is a no cost or low cost solution, or partial solution, to the imbalance.
- If you see that a line is not in balance, then a rebalancing exercise will smooth out the flow and the line will improve its KPIs. Double handling will reduce and productivity will generally increase if there is a steady flow in production.

Staff not appearing to have anything to do

- This tool in the box is one that has to be handled very carefully.
- "Staff with time to lean have time to clean."
- If there are staff who are routinely not able to do their own job and are used elsewhere in the business, it is an indication of an opportunity to improve performance.

The reallocation of staff and the redistribution of their role to others is a common way of ensuring that they do have sufficient work to keep them occupied. Often by looking carefully at the work that needs to be done, it is possible to design the perfect system on paper. It is just then a matter of applying that system in practice.

One useful tool here is the concept of Takt time. "Takt" is a German word for rhythm or beat. The Takt time of a department is like the heart rate or clock speed that is required to allow the department to meet its output requirements. For example, a packing department has to produce 20,000 packs in an 8 hour shift. The Takt time for the department is therefore 1.44 seconds. The department has to make 1 pack every 1.44 seconds to meet its target of 20,000 in an 8 hour shift. (8 hours is 480 minutes, which is 28,800 seconds). The ticking rate of the department is 1.44 seconds. All tasks in the department should have the correct quantity of resources to meet that Takt time. Too little resource and the machine will stop. Too much resource and there will be idle time in the people and machines.

An investigation and improvement activity into finding the required Takt time for an area is a first step in looking for opportunities to improve performance (see Figure 11.33).

The views of visitors to your factory

The final idea of spotting opportunities to improve is easy to do, but among the most difficult to do well. People will visit your factory or your area on a regular basis. These people are a source of ideas and they will often spot an opportunity that you have walked by for months, even years:

Figure 11.33 The Takt time of a production process is like a metronome or drum beat. With every tick there needs to be another pack made. If sections of the line are running faster than Takt time, it indicates they will have idle time soon. They will also be causing waste by building stocks downstream. If sections are running slower than Takt time, it indicates that stock will be building upstream of the operation as they fail to keep the required pace. Takt time does not give you the answers but it is a very good way of looking at what can be a very confusing picture of fast pace production in a food factory situation.

- Your customers will automatically feed back improvement "opportunities". They will see ways in which you could improve. They will not always tell you it is a way to reduce cost, but their suggestions often are. A complaint about product quality will be an opportunity for you to get the product right first time, so that rejects occur less often. A complaint about the department looking untidy is an opportunity for you to improve hygiene, safety and efficiency at the same time. Is it untidy with waste? Is it untidy with unused machinery? Is it untidy with product waiting to be fed into a machine and therefore being double handled?

But all visitors will do the same:

- They will rate your factory in comparison with others that they visit, if you ask them to. How about developing a visitor feedback form to capture this valuable information from them. Once they have left the factory, the opportunity to use their knowledge has left as well.

- Pest controllers will get to all corners of your site and will offer their views if asked. It might only take one of the team to spend 10 minutes on a debrief at the end of their visit to gain a list of things that could be improved.
- Contractors will let you know if your business is up to scratch. Service engineers are experts on the machines that you are using. While they are on site to replace that circuit board or load a software upgrade, get them to look at that infeed section you have noticed is generating more waste than it used to.
- Agency and temporary workers are a great source of opportunities to improve, if you ask them. Agency and temporary workers visit more factories in a month than you might in your whole career. There will be ideas in their head that they would share with you for the price of a cup of tea.

This concludes the section on collecting information and spotting opportunities to improve. These are two of the key skills in Lean Manufacturing, and as many people as possible in a factory need to understand and use those skills.

The improvement cycle

Within Lean Manufacturing there is a need to make improvements continually in order to keep the performance of your factory moving forward. Each improvement takes place in a series if stages and these are often referred to as the Process Improvement Cycle (see Figure 11.34).

Improvements are spotted first. They are then investigated and a solution, or possible solution, is planned. Next is the Do stage where the plan is tried in a small way to see if it will work. Often product from that trial, if any, will be discarded as it may have been produced outside of the required specification. The food manufacturing industry needs to be carefully controlled to make sure food safety is not compromised with any change, so the plan is next checked for all possible knock-on effects. If all is well, the plan is introduced and the improvement is made.

The secret of successful organisations is the speed with which they can get around the improvement cycle. Once a month will see improvement activity being introduced very slowly, once a week is better but quite slow in an industry that is changing so rapidly. Once a shift would be a high level of improvement activity. Some leading companies that use Lean Manufacturing techniques have a target of an average of twice round the improvement cycle per shift. They even have a KPI to measure the average rate of improvement activity in the business.

The financial effects of eliminating waste

By definition, waste costs money, so when waste in all its forms is reduced, that cost is reduced but the full impact of eliminating waste from a business is much greater than the financial one. The financial one is probably the most important one in most businesses.

Here is a summary of the costs of running a food business for a week (see Figure 11.35).

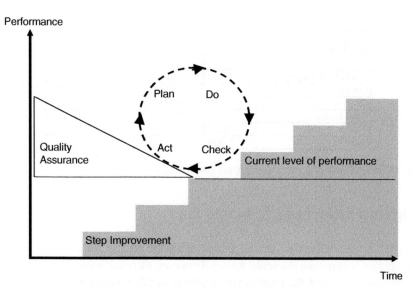

Figure 11.34   Performance increase over time as teams of people work their way around the improvement cycle. Plan – Do – Check – Act – is a typical cycle.

Cost and Profit Summary		
	£	£
Sales Value	85045	
Cost of Food Materials		21547
Cost of Packaging		16268
Cost of Labour		28145
Other Costs		15847
Total Costs		801807
Profit	32.38	
Profit %	3.80%	

Figure 11.35   Cost and Profit Summary.

This shows that the costs of the business are £81,807 and the profit made in the week is £3238 or 3.8%. It would be interesting to see what would occur if the team in the factory focusing on the seven wastes were able to reduce the costs by just 5% for food materials, packaging and labour. Assume that the sales and the other costs are exactly the same at £85,045 and £15,847 for the week (see Figure 11.36).

This shows the power of reducing waste within a business. A 5% reduction in the cost of materials and labour has resulted in the profit doubling to £6536 for the week. All the savings made are said to have "gone straight to the bottom line" and increased the profit for the business. If the factory performance remained the same, the only other way to double the profit would be to double the sales value. The factory would have to produce twice the volume. It is quite common for a company to announce that their profit has doubled since last year. That is nearly always following a period where the business has focused on the seven wastes and started working differently.

Cost and Profit Summary							
	£		£		£		£
Sales Value	85045				85045		
Cost of Food Materials			21547				20470
Cost of Packaging			16268				15455
Cost of Labour			28145				26738
Other Costs			15847				15847
Total Costs			801807				78509
Profit		32.38			6536		
Profit %		3.80%			7.70%		

Figure 11.36   Cost savings and their impact on profitability.

Designing processes to waste less

A book of this type will not contain all the answers to the question of machinery and process design that will reduce the seven wastes. However, there are some principles that can be learned to ensure that the process you design is as efficient as possible from a waste point of view. These design principles were looked at in the section on opportunity spotting and information but, to act as a summary, here they are again.

One touch processes will be more efficient from a material and labour point of view. Stocks should be reduced or preferably eliminated. This will move the factory towards a JIT production process. The factory should produce exactly what is planned, not under- or overproduce. Products must be right first time to reduce rejections and inefficient rework. Production lines and processes must be balanced and the line should not run faster than the slowest part. It goes without saying that a line should be designed to minimise product losses through product falling on the floor or becoming contaminated in some way.

Energy and water usage

There is one cost in the food manufacturing industry that is not always obvious from the point of view of people working on the shop-floor. Recent changes in the cost of energy and the realisation that water is not an infinite resource has lead an increasing number of food companies to focus on the cost of energy and water in their business.

Food production is a process that can be very energy and water hungry. Vast quantities of energy are required to cook and cool foods. Hygiene and washing procedures use large volumes of water. Waste of energy and water can be very costly to a company in its utility bills but it will also harm the business reputation and its image as a responsible manufacturer if energy and water consumption are not optimised.

If you want to manage something

How can a company save cost in this area? "You cannot manage it until you measure it." The key first step in the control and reduction step is to measure what is going on.

Daily or even hourly meter readings will tell you at what time the energy and water are consumed. The data will also point you towards areas for improvement. Better still is to sub-meter the supply, so you can collect data on which department or even which machine is generating the biggest cost. Some analysis of the data will then point to possible opportunities to reduce cost and increase profit.

In the area of energy and water consumption, there are some technological improvements that would reduce cost if an investment could be made. The best option is nearly always to go with the low cost/no cost options first. These quick fixes are often simple measures that will have an impact on the costs and also will act as a motivation for the people and teams doing the improvement actions. Simple things, such as turning equipment off when not required and repairing leaks, are solutions that can and should be done today. One of the technological innovations is the use of stepper motors rather than compressed air to operate machinery. Compressed air is an easy to use system of energy in food factories but is very inefficient from an energy point of view. Removing compressed air systems from a machine and replacing them with the modern electronic equivalent will sometime reduce the power required to do that job by 50%.

Simpler still is a technique called power factor correction. This reduced the size of an electricity bill for a factory by ensuring that the equipment is controlled in a way that reduces the peak power demand for the factory. Electricity bills for factories are based not just on the quantity of power consumed but also on the peak demand. Power factor correction achieves this by ensuring that the three phases in a three-phase power supply are kept in balance. Water consumption in a washing system can be greatly reduced by using rinse water from one cycle of the machine to carry out the pre-wash of the next cycle. The water is used twice before it is sent to the effluent plant. The list of possibilities in the area of energy and water consumption is almost endless, but as always the first step of any project to improve efficiency in this area is to measure.

For example, in order to measure and control water usage in the various departments of a food factory and on the production lines, each section of the factory was given its own water supply. Once accurate measurements could be taken, it was possible to manage the water consumption and make improvements. Without the meters, small improvements would have gone undetected and as a result no one would know if an improvement idea had worked or not. This company made a 28% reduction in water consumption in the first year after installation of the meters, which paid for themselves in less than 8 months through reduced water bills and effluent charges.

The year after they did the same thing for the electricity supply (see Figure 11.37).

Questions

- Does your factory look into the area of water and energy waste?
- Think of three examples of something that might be looked at to reduce consumption of energy or water.

Quality management

The description of the seven wastes has shown that processing waste and reworking are two key areas of focus. The management of quality in a Lean Manufacturing food factory

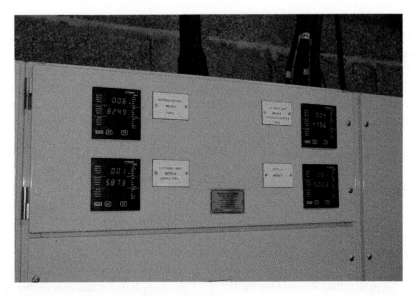

Figure 11.37   Electricity consumption was broken down into four sections, so that each could be monitored, controlled and improved separately.

is vital to the success of the business. It is vital that the quality of the products in the factory is assured as the materials flow through the process. "Right First Time – Every Time" is the philosophy that needs to be followed, but in the event of a fault developing in the process, it is vital that this is detected quickly and that materials with critical faults are not allowed to proceed in the process and are rejected as soon as possible.

In a Lean Manufacturing food factory, the measuring and assuring of the quality of the products and processes takes place at every stage. The worse possible thing that could happen is to allow a faulty product to be made and then have to reject it or rework it to bring it into specification.

The first stage in a quality management system is to agree with the customer the quality criteria and the target values that are going to be used to judge the product and then to ensure that those criteria are measured as early in the process as possible. On a loaf of bread, the colour of the crust may be one of the criteria and a target colour needs to be agreed with a maximum and minimum value set. That colour should be measured in the factory as the bread is coming out of the oven. If it is measured later, say after the bread is cooled down and is about to be sliced, there would have been a 2-hour delay and 2 hours worth of reject product could have been manufactured.

The summary of the role of quality management in a Lean factory is to measure early and take corrective action quickly. This can be achieved in many ways, but in most Lean factories the task of measuring and taking action is carried out by production operatives. The operatives are trained in the standards required and able to take the required action to correct the situation and prevent product from drifting outside of the specification. Of course, in well-managed factories, it is the process that is managed rather than the product. The majority of the quality of the products is assured through a system of rigorous process

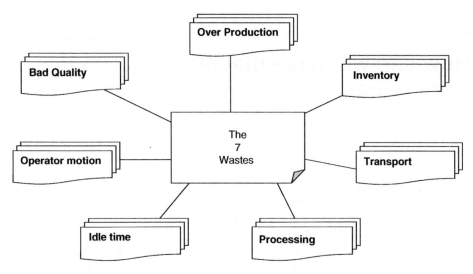

Figure 11.38   The seven wastes that need to be eliminated or minimised through Lean Manufacturing techniques.

control. SOPs are used to ensure that a standard process is created for the materials to flow through. By monitoring and controlling the process tightly, the products are consistently of the required specification (see Figure 11.38).

Summary – The seven wastes in the food industry

Lean Manufacturing is essentially the reduction of unnecessary cost, so a study of the seven wastes and the ways to spot them in your factory is essential if you are going to improve performance.

First, waste is not just that in the skip at the back of the factory. These seven forms of waste are all damaging to your business and profit margins at all times. Some of the waste is difficult to spot and some is difficult to avoid, but Lean factories minimise it as much as possible.

In identifying and analysing some of the forms of waste, the concept of a Takt time was introduced, an essential rhythm the factory needs to operate to. Anything that is not working to that rhythm is a cause of waste. Finally, by looking at the concept of quality management in a food business, the conclusion is that the best way of managing quality is to ensure that quality issues are raised as soon as possible in the process, so that no further cost is spent on material that will not make the grade. Sweeping quality issues under the carpet is in itself a waste, as every quality issue is an opportunity to improve, an opportunity to get it every time.

Chapter 12

How can we make machines work better?

Food factories are an interesting mix of people and machines brought together to achieve demanding tasks, often under time pressures, in the manufacture of products with low profit margins. The performance of the machines in this environment often causes issues of poor performance. In a food factory operating Lean Manufacturing systems, this is obviously an area that has received a lot of attention over the years. If the performance of the machines can be increased, then the factory will be more efficient and could make more profit. But how can the machines work better?

There are areas covered elsewhere in this book that focus on skills training for the operators of the machine. There are also several systems that focus on the machinery rather than the operator. Here are some of the systems that are commonly used to improve the performance of a machine (see Figure 12.1).

Total Productive Maintenance (TPM)

This is a system that can be known by several names; maybe your factory has its own terminology for the system and may operate it with slightly different criteria. Sometimes called Machinery Care, Autonomous Maintenance or Auto Maintenance, the system has the aim of achieving zero errors, zero work-related accidents and zero losses. TPM is a system of deterioration prevention and maintenance reduction, not finally fixing machines when they are broken. The difference between TPM and normal Preventative Maintenance systems is that the work of looking after the machine is carried out by the machine operator or the production team and not by an engineer or maintenance fitter. In TPM, the machine operator carries out most, and in some factories all, of the preventative maintenance jobs. This ensures that the person with a vested interest in the performance of the machine is the one who is trained in, and carries out, the routine inspections, adjustments and maintenance tasks.

The aim of TPM, and other similar schemes, is to improve performance by reducing six types of losses on the machines.

Handbook of Lean Manufacturing in the Food Industry, First Edition. Michael Dudbridge.
© 2011 Blackwell Publishing Ltd. Published 2011 by Blackwell Publishing Ltd.

Figure 12.1 To improve anything, you first have to measure it. It is the initial performance data combined with a target that provides the motivation to change and improve. As with most things, improvement starts with knowing where you are now and knowing where you are aiming for. It is like a road map that is of no use to you if you have no idea where you are, and also of no use if you know where you are at the moment but you have no idea where you are going (picture courtesy of Ishida Europe).

Equipment failure

When a machine stops working, it is immediately apparent to everyone that a breakdown has occurred. Maintaining and adjusting of a machine will increase its reliability. As with most things in Lean Manufacturing, if you want to manage something you first have to measure it. In this case a calculation is made to work out the MTBF. The MTBF can then be monitored, controlled and even benchmarked to ensure that the TPM system is working effectively. MTBF is a common KPI for automated food factories, but there are other ways that performance in this area can be monitored and managed. One that is often used is a part of the OEE measurement described earlier. Remember availability, performance and quality losses. Equipment failure is an availability loss. The machine needed to be run but could not because it was not working. You will remember that not all availability losses are due to machine breakdown, but all machine breakdowns are availability losses (see Figure 12.2).

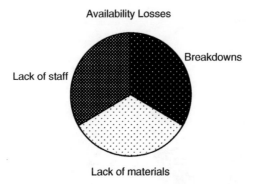

Figure 12.2   Machine availability losses: TPM is focused on maximising the availability and performance of food production machinery. TPM would have no impact on the availability of materials for the machine or of the availability of staff.

Reduced speed

This is often less obvious. A machine is running slightly slower than it should. Often this is not noticed until the end of the shift when it becomes apparent that the expected output has not been achieved. Again Lean Manufacturing has a system to ensure that machine speed is monitored. SICs allow the production team to constantly check where they are against the planned output and to take corrective action if they are falling behind plan. Reduced speed commonly occurs in two different forms on a piece of food production machinery. The first is that the machine is obviously running slower. The machine is running well but just at below the target speed for that product. This could be because someone has altered the controls to make the machine slow down or it could be for a mechanical reason. A drive belt may be slipping or a pneumatic cylinder may be operating slowly because of reduced air pressure. There are many possible causes.

The second reason is that the machine is running at the correct speed but that every few seconds it "misses one". This is commonly seen on flow wrapping machines and cartoning machines, where the machine is running but it fails to make a product once in a while.

For example, a flow wrapping machine was running at the set speed of 100 packs per minute and tests with a stop watch showed that the speed indicator was correct. However, it became apparent that only 92 packs per minute were arriving in the boxing area. The flow wrapping machine was fed by a flighted infeed conveyor and one of the flights had been lost and as a result the performance of the machine had dropped by 8%.

As soon as the lost flight was replaced, the machine output returned to 100 packs per minute.

Set-up and adjustment

TPM has a role to play in ensuring that a machine can be set up accurately and be running and making good product as quickly as possible. If the adjusting mechanisms on the machine are dirty or worn, the machine will be difficult to set up and adjust. Making sure

Figure 12.3 Even small variations in angles and temperatures on high-speed wrapping machines can have a huge impact on the performance. Careful adjustment and precise calibrations are required to make these machines work well from the first minute of the shift (picture courtesy of Ishida Europe).

that all adjustable parts are clearly marked with the required positions and settings is part of visual control that was covered earlier in the handbook. The setting and maintaining of the optimum positions of movable parts is part of TPM (see Figure 12.3).

Idling and minor stoppages

The machine is running but every now and then a pack gets jammed or a reject is produced. In order to clear the blockage or make a minor adjustment, the machine has to be stopped for a few seconds. While this is not a breakdown, the effect of repeated minor stoppages, sometimes called short stops, can add up to considerable lost time on the machine. Another issue with short stops is that when the machine is restarted it is not uncommon for the first few products made to be rejected as they do not meet the required standard. These reject products will end up as waste or be reworked and the effect has been seen in the section on the seven wastes. Idling of machines refers to a machine that is running but not producing products for some reason. This can cause the conditions inside the machine to change so that when it starts making product again, the first products made can be faulty in some way.

For example, on a flow wrapping machine, if no product is fed to the machine, it will idle. When product is refed, the machine will start up automatically, but by then the temperature of the sealing jaws may have got too high or the date coding system may have become misaligned. A similar thing happens on cartoning and boxing machines; often after a period of idling or a short stop, the glue system fails to operate correctly and a carton is made without glue. TPM can assist in ensuring that the machine gets it right first time, even after a period of idle time or a short stop.

Reduced yield

TPM will have a contribution to the performance of the machine in terms of its yield during normal running. A correctly maintained machine will have less variability and as a result produce more of its products within specification. As a result of the tighter control over the variability, the machine can to be adjusted to get closer to the ideal settings to maximise yield. The other area where yield should be considered is in the period immediately after start-up or restart from a short stop. A well-maintained machine will produce less out of specification products than a machine that has been poorly maintained. The chapter on product costs showed that the food materials used in food factories make up a substantial part of the overall costs of the business. Having machinery that can operate in a tight window of specification will allow a skilled operator to maximise the yield at the same time as making a consistent product.

Product rejects

A machine that is subjected to a correctly running system of TPM will produce less rejects and minor defects and as a result, quality and waste levels will both improve. It is an obvious statement that keeping machines running at peak performance will reduce the overall cost to the business. Some cost will have to be expended to get the high level of performance, but the rewards in increased output, consistent quality, improved productivity and high levels of on-time, in-full delivery usually pay for the additional cost of a TPM system.

The aim of a TPM scheme is to maintain machinery in "show room condition". That is, to ensure that the machine performs as if it were brand new.

TPM is a system of maintenance in a factory that ensures that machines are performing at their best, and as a result the costs of the business associated with machine performance improve. TPM requires an investment in time for the operator but if correctly managed, TPM systems can improve productivity and reduce operating costs in food factories.

If TPM is about maintaining a machine in its original design condition, then the next way of making a machine work better is about improving that design (see Figure 12.4).

TPM is a powerful tool in the race to improve performance. What do you think are the likely issues that have to be overcome to start using TPM type techniques?

- People
- Machines
- Safety

Everything that prevents TPM techniques being used can be addressed with some investment in the people working in the factory. The training of people can be achieved, as can the modification of machines, to make safe access easier and unsafe access

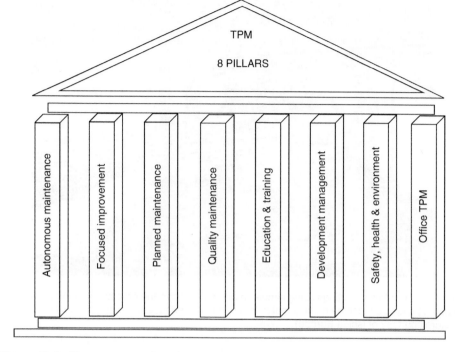

Figure 12.4 There are several versions of TPM that are used by different companies but all have common themes as a manufacturing philosophy. This model shows eight pillars and covers all major points that need to be considered. Without paying attention to each of the pillars, the house becomes unstable and may fall down.

impossible during the TPM routines. One of the main stumbling blocks for TPM tends to be time. More people are often needed in production to release the people carrying out the TPM work. The hope, of course, is that the increase in performance outweighs the extra investment.

Continuous improvement and focused minor modifications

Food processing and packaging machines are often designed by engineers who have never worked in a food factory. It is possible for your company to specify a machine to a machinery supplier, but often the specification will not be complete and the machine will not be optimised for the job it has to do. The machine is delivered that appears absolutely ideal for the product and process to be done. The food industry changes so fast, with both products and processes being updated on a regular basis, so within a few weeks, or months, our machine is no longer optimal.

There are many machines used in the food industry that could have their design modified slightly to improve some aspect of their performance. That modification could take the form of a software change to adjust the machines output or could be a change to a

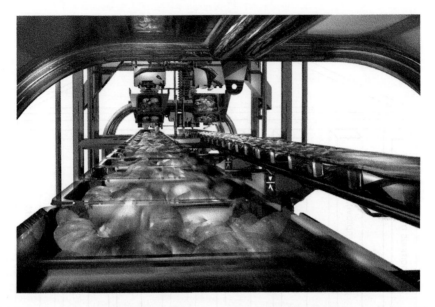

Figure 12.5   Making settings using simple markers can make the process of running and changing over the machine much simpler. In a complex food factory, simple is usually good (picture courtesy of Ishida Europe).

motor cover to improve cleanability. This can be something as simple as changing all the bolts on the machine, so that the operator only needs one spanner to carry out all TPM requirements, or changing a fixing from a bolt to a quick release clip to make changeovers quicker. How about a modification to stop water ingress during a hygiene routine or another to mark settings on the machine in different colours for each of the products that is to be made. Twelve-inch pizzas could have the red settings and change parts. Eight-inch pizzas could have blue (see Figure 12.5).

Improving on the equipment design

Machines can be made to work better once an opportunity has been spotted. The system of teamwork and continuous improvements allows a machine to be improved to make it better than the original design. This means that the machine performs at a higher level in some way as a result of the modification. Any of the six losses associated with the use of machinery can potentially be addressed in this way.

In carrying out machine modifications, it is vital that expert help is brought in. Modifications may improve the speed of the machine, but there may be a price to pay in terms of reliability or accuracy. Modifications may make the changeovers easier and quicker, but can compromise the safe operation of the machine. The help required to make modifications may be available from your own engineering team, but it is worth getting the original equipment manufacturer involved as well.

In the improvement of machine output and performance, it is also worth remembering that the machine can only improve its output if the process feeding the machine and

Figure 12.6 Three phase error proof connections.

the process taking the product away also increase. The line balance needs to be maintained, or an improvement in one area will result in an increase in one of the seven wastes elsewhere.

One method of modifying machines, and a general principle of Lean Manufacturing, is the elimination of errors.

Error proofing

One of the causes of problems with machinery under performance is that the operator of the machine has made an error. The wrapping film was threaded the wrong way, the wrong plate was fitted at the last changeover, the machine settings were wrong, or that adjuster was in the wrong place. Now, the reduction in errors can be addressed with additional training but that does not guarantee that errors will not occur again. There are techniques used as part of Lean Manufacturing that effectively error proof a situation, so that the only way to do something is the correct way, or if that is not practicable then the next best error proofing method is that it becomes immediately obvious to the operator that an error has been made.

Some examples of error proofing will already be in place in your factory. Try to plug an electrical machine into the wrong type of electrical outlet! Colour coding and size differences make it impossible to plug a machine that requires a 110 v supply (yellow socket and plug) into a 3 phase 415 v socket (red, 5 pin and bigger) (see Figures 12.6 and 12.7). A lot of error proofing is usually built into safety critical parts of machines. Try to open the safety guard on a machine; the machine stops.

Knowing that errors cost us time and money in our factory, why are there less systems of error proofing on performance critical parts of our machines. In fact, there probably are some. A part of a machine that is regularly adjusted during changeovers will usually have a scale or mark system to aid the correct positioning of the part. Change parts often have fixing bolts that only allow the change part to be fitted in the correct way.

Figure 12.7   A socket for power supply: colour coded, shape and pin size all protect against errors.

Error proofing comes in three different categories.

First, there are *warnings* that give a signal of a potential problem, such a red light on a control panel or a buzzer to attract an operator to a control panel.

Next are systems that do not prevent an error but will *stop* a machine if an error is present. For example, there could be an auto-coding system to check that the date code on the pack is the correct one.

Next on the error proofing scale are systems that prevent or reduce the chances of an error being made by a skilled operator. These are systems such as components that only fit one way, or colour coding to prevent similar components or materials being mistaken.

The final system of error proofing is one that involves auto-correction or the error. Systems of checking that top and bottom labels on a pack are for the same product or that the outer case label matches what is inside the box. The checking system can only be overridden with the password of the departmental manager.

By having a machine that is looked after using a system of TPM, it is optimised in its performance using minor modifications and continuous improvement and the number of operator errors is reduced through error proofing techniques. What else can be done to maximise the performance of the machine?

- Think of three examples of error proof systems in your factory.
- How about one that you would like to see?
- What was the last mistake made in your area?
- How could that be made error proof?

Quick changeovers

Techniques for quick changeovers were developed in the engineering industry, so are sometimes called Single Minute Exchange of Dies (SMED).

It is a characteristic of the food industry, because of the demand for variety and need to keep finished product stocks as low as possible, that machines have to be flexible and are changed over on a regular basis from one product to another. It could be a simple flavour change or it could be from a 12 pack to a 6 pack. It could be a product of the same format or could be something totally different.

The need to change a machine from doing one task to another may occur once or twice in a week in a factory making long shelf-life foods, but in a factory making short shelf-life fresh foods, changeovers could be taking place once or twice every hour.

The need to change over as quickly as possible is the same for both factories. Changeover time is non-productive and as a result has a big impact on the costs of the factory. The time it takes to change over is classed as the time from the last good one of the original product to the first good one of the new product.

Rapid changeovers are sometimes called pit stop changes. They require the same quality and quantity of planning and it is essential that everyone on the crew knows their role in the pit stop. It is a sure thing that the first time a racing crew tried to change a set of wheels in seven seconds they did not manage it. They will have studied the task, developed new tools and organised each role until the time came down to the target.

In order to reduce changeover time, Lean Manufacturing systems have a method to follow and as with all improvements, the first step is to measure.

There are six steps to quicker changeovers.

Study the current changeover process

Observe and record the current changeover process and record all tasks that have to be completed for the changeover to be successful. It is possible to time the individual parts of the changeover to allow for analysis later. The observation should be of the entire changeover, not just the bit while the machine is stopped.

Internal or external tasks

Label each of the tasks in the changeover process as external or internal. An external task is one that takes place while the machine is still running, so it could be fetching the new batch of raw materials from the store, or getting a bucket of water ready to wash out the machine (or indeed, taking the bucket back once the new product is running). Internal tasks are those that take place once the machine is stopped. These could be changing machine settings or swapping changeparts.

Convert internal to external

The next step is to analyse the tasks and convert as many as possible to external tasks that can be carried out while the machine is running. These tasks can be carried out both before the machine has stopped and after it has restarted. External tasks are not idle time. The internal tasks those inside the period when the machine is idle. To minimise the idle time and get back into production, it is the internal tasks that need to be slick.

Make the internal tasks efficient

Now it is time to focus on the remaining internal tasks and make them as efficient as possible. This can be done using various methods of reducing the time taken. A simple one is to have more than one person working on the changeover. Imagine how long it would take a Formula one team to do a pit stop if just one person had to do all tasks. The pit crew is well drilled and the tasks are analysed in great detail. This is the sort of approach needed to improve changeovers in your factory.

If the changeover needs something to be unbolted, can the bolt be replaced with a quick release system or if not, can the bolt be made shorter so it requires less turns? If the changeover requires that something is cleaned, can this be organised to happen very quickly or can a second component be purchased so that the dirty component can be swapped and the cleaning then becomes an external task.

Focus on the set-up time to first good product

Once the changeover is complete and the machine restarts, how long is it before it makes its first good product. Does the machine require "setting up" or adjusting? The aim of SMED is to make sure the first product off of the machine is a good product, so there is no room for setting up. Parts of the machine that are adjustable should be put to the required standard setting during the changeover by using gauges and indicators. If a machine needs to warm up after a period where it is stopped, then try and find a way of preheating the component to reduce the warm-up time. During a Formula one pit stop, the tyres are prewarmed when they go onto the car to ensure that they are up to racing temperature as soon as possible. The same should be true of a heat sealing die or the jaws of a flow wrap sealing machine (see Figures 12.8 and 12.9).

Make the external tasks efficient

The final element is to re-examine the external tasks and make them as efficient as possible, to reduce the overall cost of the changeover.

Now the machine is close to optimal performance, the final step is to keep it there using process control, to ensure that the machine is doing what it is supposed to do.

Statistical process control

A food process is difficult to control for many reasons, not least of all because the raw materials fed into the process are natural and have an inherent variability. Because of this natural variation in a process, it is often difficult to determine if the process needs adjusting or if the change is a result of a natural variation and will come right by itself. There are techniques that can be used to predict the situation.

The aim of Statistical Process Control (SPC) is to create information that will aid decision-making in the factory. As you would expect, SPC looks at data and compares it to targets or set points and works out what the correct response should be. The response decision is based on a knowledge of the process and its inherent variability.

SMED-Changeover analysis

Changeover

	Sauce depositor		
Task	External or Internal	Time taken (sec)	Number of operative
Switch off line	Internal	1	1
Unplug air line	Internal	5	1
Pull away from production line	Internal	5	1
Empty original sauce by hand	Internal	120	1
Unclip hopper	Internal	10	1
Carry hopper to was area	Internal	30	1
Wash hopper	Internal	480	1
Take deposit head off of machine	Internal	60	1
Carry deposit head to wash area	Internal	30	1
Wash deposit head	Internal	600	1
Return clean deposit head to line	Internal	30	1
Return clean hopper to line	Internal	30	1
Reassemble machine	Internal	120	1
Push depositor onto line	Internal	5	1
Reattach air line	Internal	5	1
Fill with new flavour sauce	Internal	30	1
Switch on line	Internal	1	1
Total		1562	

Figure 12.8 It can be seen here what happens when the team leader is the only member of the team with the training to carry out a changeover. All tasks are completed by one person and all tasks follow on one after the other, giving a total time of 26 minutes to change the flavour of sauce in a depositor. Working as a team and using the techniques of SMED, the changeover could be reduced considerably. In this case, a second depositor standing by with the new flavour of sauce would convert all of these tasks to external and the changeover time to zero. There would be no need to stop the machine; just switch off one depositor and switch on the next. From strawberry jam tarts to lemon curd, without missing a beat!

SPC is a specialist area and beyond the scope of this book, but the use of SPC techniques is widespread in the food industry, so it is worth have a look at some to the systems that you may have in your factory.

Control charts

These are graphs that are built up during a shift and will record the history of a particular variable. For example, the diameter of the biscuits at the end of the baking process varies during the shift. Some of the variation is the result of the natural ingredients but some is caused by slight changes in the process. How does the person on the end of the line know if the change will correct itself or if a change in the process is required to alter the diameter?

One type of chart that is used in these circumstances is called a Shewhart Chart. A Shewhart chart contains a target line together with lines that are called the upper and lower warning limits and the upper and lower action limits. If a data point plotted on the chart falls outside of the warning limits, another check is carried out. If that too is outside of the warning limit, then the operator needs to make an adjustment to the process. If a

Figure 12.9   Changing over quickly on this meat packing machine could be the answer to a rise in lost time on the line (picture courtesy of Ishida Europe).

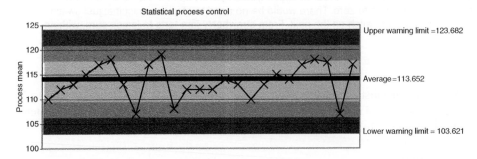

Figure 12.10   On this Shewhart chart the CENTRAL zone indicates the product is under control and all is well. The OUTER zone is the warning zone. Data in this zone could be as a result of natural variation so if an adjustment is made too quickly the product could move from one extreme to the other very rapidly. The AREA BEYOND THE OUTER zone is the Action zone (sometimes called the Upper Control Limit) a data point in this zone will not have been caused by natural variation alone and corrective action should be taken immediately.

data point falls outside of the action limit, then adjustments must be carried out on the process (see Figure 12.10).

In our example, the diameter of a Shewhart chart can help with the control of the process in keeping the products within specification, even when the raw materials from which they are made vary (see Figure 12.11).

Figure 12.11 Salt sticks from a production line. Control is vital and SPC can help (picture courtesy of Ishida Europe).

SPC systems are now built into modern check weighing systems and data is automatically plotted and displayed. Some check weighers even feed the information back up the production line, so that corrective action is taken automatically.

Summary – How can we make machines work better?

It is often frustrating for people working in food manufacturing that the performance of the machines appears to be far from perfect and that waste is created as a result; waste of materials, waste of manpower, waste of time, waste in all of the seven ways that have been identified.

Lean Manufacturing has techniques that can help improve this situation, such as the few already described in this chapter.

The first was Total Productive Maintenance (TPM), a large group of techniques that are aimed solely at getting better performance out of a machine. This chapter showed the main features of this group of techniques, and that keeping the machine well maintained is number one on the list.

Knowing the reasons why production lines need to be changed over on a regular basis, the concept of SMED was introduced; rapid changeover techniques. SMED is able to increase the availability of a machine for work, by changing it over in the minimum amount of time possible.

Finally, the possible impact of SPC is an aid to decision-making on the shop-floor. In using SPC techniques such as the Shewhart chart, it is possible to detect the difference between natural, and expected, variation in the process and a variation that needs an adjustment to keep the process under control, and to keep the products within specification.

Chapter 13

How can we let people contribute more?

People are a valuable asset, especially in the fast changing world of food manufacturing. People have flexibility, an ability to learn, can make decisions and think, but most importantly, can have ideas.

Who can give more?

The simple answer here is that everyone can contribute more to the business. The best ideas can come from anywhere within, or even outside, the business. There are no boundaries to inspiration. The important thing here is that the systems within a food factory are set up to make sure that when an idea surfaces it can be captured. It could be that the idea is rubbish, one with not a hope of success, but the person who came up with it needs to be encouraged to come up with another one, not be made to feel inadequate.

It only takes one idea to improve the waste on a production machine, one idea that can improve the energy consumption or packaging usage. What a food factory that is practising Lean Manufacturing needs is a thousand new ideas and a way of picking the winners (see Figure 13.1)!

Work smarter not harder

Often it is said that Lean Manufacturing is about working smarter and not harder. This is true and that requires brainpower from the people in the factory. Can you imagine the effect in a factory if everyone was working smarter? But what about the effect in the factory if everyone is working smarter *and* harder. Generally if people are listened to and their ideas treated with respect, they will tend to work harder as well as work smarter. This occurs as a result of them feeling that they are part of the team and that their views are taken into account. The food manufacturing industry is a tough one, and operating to high levels of performance and low profit margins makes it difficult to sustain and grow a business. The use of Lean Management techniques do allow the, often conflicting, pressures in the industry to be balanced.

Handbook of Lean Manufacturing in the Food Industry, First Edition. Michael Dudbridge.
© 2011 Blackwell Publishing Ltd. Published 2011 by Blackwell Publishing Ltd.

Figure 13.1   This production line could improve its performance based on just one idea. The techniques of Lean Manufacturing make sure that plenty of ideas are generated by making sure that everyone has the information needed and the motivation and opportunity to suggest changes. The people on this line may have many ideas that individually will have a small impact, but when added together they can make a huge difference to the operation of the line.

Other sources of ideas and brainpower

By focusing on people, it is possible to get brainpower for free. Employees all come with an ability to contribute, even though their current role is very limited. A survey of employees in a food factory will always reveal special skills and knowledge, interests and motivations of an employee that could be used to the benefit of the business and also enrich the role of the employee.

Who better to be part of a group looking at speeding up changeover times on a machine than someone with an interest in Formula one racing? Pit stop wheel changes are the very essence of quick changeovers and the techniques used by the Formula one teams could be applied to a changeover in the factory.

Who better to be part of a team looking at innovative new products than someone with an interest in food and cooking? If it is felt that an area is in need of some reorganisation and an upgrade in 5S, there will be someone in the team who is passionate about things being tidy and organised. Get them to lead the improvement team.

Employees are not the only source; what about the employees' families and friends? A well-motivated employee may talk about an issue in the factory with their partner or friend, who happens to know a bit about that issue. There is no shortage of ideas but there is a shortage of ideas that have been created from a position of full knowledge about the background to an opportunity. For that reason, it is vital that information about the performance of a factory is widely known in the workforce. The workers need to know what the main issues are and what the constraints are for possible solutions.

Brainpower can also come from the local government officers, trading standard and environmental health, or the fire brigade and police. How about local schools with a need for food technology projects; who better to help your new product development process for a product aimed at them. Local universities are often able to provide an injection of brainpower; how about a tour of your factory in return for some ideas about an issue that you have.

Setting development plans

The difficulty of a system that generates thousands of ideas (hopefully) is that it will be very difficult to keep focused, so the process needs to be organised. This is often done with a system of Personal Development Plans (PDP).

PDP schemes work best if the objectives of the business are set and communicated first and then individuals can suggest ways in which they could contribute to that overall plan. A business may have detected in a benchmarking exercise that it needs to improve its use of energy and sets a target of reducing energy consumption per tonne of product by 20%. Individuals in the business need to identify in their PDP what they can do to contribute to this overall aim. The managers control and agree who is going to do what and put teams together to look at particular aspects. The outcome for our factory might be that a team leader with an interest in Formula one racing is on a team looking at changeover times and the resulting increase in output, and our home cook is going to come up with some ideas of products that require a shorter cooking process.

PDPs will also allow the business to focus on training of its people to build knowledge and experience.

- How are PDPs used in your business?
- When did you last think about your own development or the development of those around you?

Training for increased performance

Food factories are complex places, with complex processes and machinery. The process of manufacturing food is one where technical standards need to be high to maintain the quality and safety of the food produced. Add to that the fact that food factories require people to work together, often under time pressure, efficiently and effectively on products that have a tight profit margin and the need for training, is easy to see.

Decisions in a factory operating Lean Manufacturing techniques obviously need to be made correctly, but they also need to be made quickly. People in the factory need to feel empowered to make decisions but also need to be trained to reduce the chance of a wrong, or even dangerous, decision being made.

This book is not able to list all of the training required to operate in a food factory. That will depend on the individual, the demands of their role and likely decisions that they will

Figure 13.2 The need for training is an obvious one in such a technically demanding industry as food manufacturing. Training should have a purpose beyond compliance with regulations and rules. Training should be used as a method of getting employees engaged in the business and equipping them to make decisions.

be involved in. However, for a food factory to perform well, training needs to be organised, planned and controlled.

A training system needs to be designed where people's training needs are assessed and training is delivered in an appropriate manner. Training can be delivered in many ways. Some factories have a training department with coordinators who ensure that the required training takes place and that training records are kept updated. Other factories rely on the production management team to run the training system.

However, the training is organised and needs to be delivered in a usable form that improves the trainee's knowledge and performance. Training should always be for improvement in performance and not simply to comply with external requirements.

For example, everyone working in a food factory needs to be trained in food hygiene, so that the risks in the factory can be minimised. The fact that external auditors want to look at the training records for hygiene training is secondary. The fact that everyone is trained in hygiene does not mean that everyone will abide by the hygiene rules. If there is widespread abuse of this vital area of food manufacture, then the training has been inadequate and maybe needs to be redesigned (see Figure 13.2).

Figure 13.3 Food factories require staff who are individually skilled but who also are able to work as part of a team to achieve shared goals. Food factories are often very complex places and there is a need for a mix of skills and personality types.

Lean Manufacturing techniques rely on people wanting to contribute their ideas and feeling empowered to make suggestions for improvements. Better training means better suggestions, and better suggestions mean faster improvement in performance.

Selecting team members

Food factories are places where people work together to achieve plans and improve processes and performance. If people are to be effective in a factory environment, it is important that they work effectively in a team.

Some factories use psychometric testing to ensure that the people they employ or place into new teams will fit with the culture of the team they are joining. The tests are a great way of predicting the performance of an individual and can be used to help plan careers and role changes (see Figure 13.3).

An effective team is a group of people who are working together. This collaboration is to reach a shared target for which they hold themselves responsible. A group of people is not necessarily a team. A team is a group of people with a high degree of interdependence geared towards the achievement of a common goal or completion of a task.

Team members are deeply committed to each other's success and personal achievement. A team outperforms a group of individuals and outperforms all reasonable expectations given to its individual members. That is, a team has a synergistic effect; one plus one equals a lot more than two.

Teams cooperate in all aspects of their roles and targets, but they also share in management type tasks, such as organising, setting performance goals, assessing the team's performance and also to develop their own systems to introduce and manage change and improvements.

An effective team has three major benefits for the factory:

1 Each member of the team is helped and supported by all the other members of the team. It is the team that feels the impact of a failure and the elation of a success. Failures are not blamed on individual team members, and that gives them the courage to take chances. Every team member is seen to be contributing to the successes, and this helps the team to set and achieve more and continue to make improvements in performance. Failure is seen by the team as an opportunity to improve.
2 There is a synergistic effect that occurs in successful teams, where the performance of the team far exceeds the performance of the individuals in it.
3 The team has all the skills, knowledge and motivation that it requires to meet the objectives it sets for itself. The performance of the team continually improves. Personal goals are put to one side to ensure that the team succeeds in its task.

With all of these benefits, it is easy to see why a factory operating Lean Manufacturing systems needs to ensure that effective teams are in place and that the teams are well put together. The task of selecting new members for a team is important, as all advantages of a strong team can be lost if the new member does not fit in and starts to disrupt the teams work. Selection of members of a team must take into account the culture of the team and its attitudes towards the task being done. A new member of a team could cause performance to drop if they had different views and were the type of person who dug their heels in over minor matters. It is impossible for us to give the full picture of teamwork and the introduction of a new team member, but the person joining the team will have values and attitudes that will need to be compatible with those of the team. For example, a team with a passion for getting the work right first time and making very few errors is joined by someone with a slapdash attitude to their work! Equally, someone who will not make decisions and prefers to stop the production process while a manager is found, joins a team that has a "can do" attitude and has been rewarded for getting the work out on time.

Total Productive Maintenance:

• Think of an example where a new team member had an immediate positive effect on performance.
• Now think of one where the opposite was true.

Distractions from the main goal

There are many possible distractions to the job of manufacturing foods. In a Lean Manufacturing environment, it is vital that these distractions are minimised, if not eliminated all together. Each individual and the team as a whole must understand the reason behind each element of the task they are carrying out. One area that is always a bone of contention among manufacturing teams is the thorny topic of "paperwork".

The food industry needs to keep records in order to ensure that the processes remain under control, technical records are maintained, and where corrective actions are needed they are recorded for future inspections. The need for paperwork in a food factory is obvious but sometimes it can be felt that the volume generated on each shift is actually stopping an individual contributing more to the business. It is vital that the paperwork completed on the shop-floor is useful and actionable. If paperwork can be minimised, the person completing it can be focused on their PDP to look at other things in the business that have some real value. The use of automatic data collection is an obvious one to look at here, but so often the auto-system is in addition to the manual system that was already there, rather than to replace that manual system.

Limits of authority and decision-making

Generally it is accepted among groups of people working in manufacturing that there is a limit to their level of authority when making decisions. The difficulty comes when it is unclear where that limit is. Teams need to have their boundaries set out for them so that they know when they have reached a point where they need to call in help from outside of the team.

For example, the team is focused on customer service level but have a problem with their packing machine. It is producing 60% reject packs, but the lorry needs to leave in an hour and if they stop the machine to fix the problem, the lorry will leave short of the full order or the lorry will leave late. The team decide to keep going and meet the order (just), but at the same time create £2000 worth of waste product.

The team did well to meet the order, but it has cost a large amount. Perhaps it should have been made clear that deciding to waste £2000 worth of product was above their authority level and they should have called the factory manager for a decision. These limits, often called escalation levels, are designed to make sure that decisions are taken at an appropriate level in the management structure of the factory.

Escalation levels of authority

It is often found that teams prefer the situation where they know exactly where the limits of their decision-making authority are. They can be comfortable that they are calling the boss at a point that they should, and they will not be criticised for not making decisions or for overstepping their authority level. Clear rules are key to the smooth operation of a complex factory, and this is no different in the food industry.

Escalation rules can be set for all areas of team decision-making. For example:

- **Waste generation** – if the waste gets above 3%, you must call the shift manager and if waste has not reduced to below 2% within an hour, you must call the factory manager.
- **Lost time** – If the lost time total exceeds 60 minutes, you must call the shift manager and if the lost time is caused by breakdowns, then the maintenance manger must attend.

A feeling of responsibility

One of the single biggest motivators for people working in a team is to feel responsible for some aspect of the performance of the team. This is one of the reasons why well constructed teams perform better than a group of individuals. Making sure that people know

Figure 13.4   The person on the right of this team looks a little outside of the group with passive body language. I would guess that a task that came up at this morning's meeting would not see him volunteering to take it on.

they are responsible and accept that responsibility is a central feature to a lot of the techniques of Lean Manufacturing. For example, the action board that is reviewed at the start of shift meeting each day has a section for "Who?" Someone's name will appear in that section and there will be no doubt in their minds or the minds of their team mates that they are responsible. It is important that the responsible person accepts that fact rather than it just be allocated to them without any consultation (see Figure 13.4).

Summary – How can we let people contribute more?

This chapter was a quick review of some of the methods commonly applied in the food manufacturing industry, to ensure that people are able and motivated to contribute at the highest level possible.

The core message is that people need clarity in their responsibilities and levels of authority. Once the situation is clear and they have a clear role and responsibilities, it is possible for them to work to their optimum performance. When roles and responsibilities are unclear, a person will avoid decisions and responsibilities on the grounds that it is not their job.

Chapter 14

How good are we?

In business it is always an advantage to be the best. The best quality, the best prices, the best customer service, the best new products, the best profit. But how do you know how good you are?

Business is a bit like a race. There is a spirit of competition between companies that makes all the companies in the competition try a little harder, think a little quicker, be a little faster, be a little cheaper. The difficulty in business is that it is not always that easy to know where you are in the race and it can be difficult to measure some of the elements being competed over.

The difficulties of comparing like with like makes the topic of benchmarking an interesting one. When looking at KPIs and the measurement of performance, it is important that every business knows what good looks like. That is surprisingly difficult to do (see Figure 14.1).

Benchmarking is a technique that allows comparisons to be made inside and outside of a business, to discover "what good looks like". A factory may believe it has the best level of performance in the country, but it cannot be sure until it compares itself with others:

- Once a good system of measurement is set up in a business, it is possible to gain even more information by a process called benchmarking.
- If measurements are consistent across several businesses, shifts or departments, then benchmarking can occur. It is important to compare like with like in a benchmarking exercise, but it is still possible to get some information from the comparisons, even if you accept that measurements are not entirely consistent.
- In benchmarking, data is shared to improve performance, by letting the business know what is possible and what is achieved elsewhere. Independent third parties, universities, etc. often carry out benchmarking exercises between companies, with the company names being kept secret. You know there is another frozen food company that has a better performance in energy cost per tonne of production; you just do not know who that is. The knowledge that you are not the best in this area could prompt you to investigate the reasons and come up with some improvement ideas.
- World Class Manufacturing is the term often given to the performance of the top 5% of companies in the world. This can be a difficult thing to decide, because businesses and industries are so different in what they are trying to achieve. The comparisons are often made in the universal measures of OEE and sometimes a measure of faults per 100,000

Handbook of Lean Manufacturing in the Food Industry, First Edition. Michael Dudbridge.
© 2011 Blackwell Publishing Ltd. Published 2011 by Blackwell Publishing Ltd.

Figure 14.1   How good is this production process compared to the competition? It is sometimes difficult to compare different factories, but it is not impossible.

products. These two measures can be used to compare the performance of radically different businesses. The real benefit of benchmarking comes from comparing your performance against others trying to do a similar thing.

- Shared information within a group is equally as valid for improvements to be made. Rather than try to source benchmarking data from competitors, it is much easier to get the data from sister companies if you are in the same group.

But all benchmarking exercises require that the measures are reasonably consistent in the businesses being compared. This can sometimes be difficult to achieve, but there are real benefits if a reliable system can be found.

Benchmarking can allow a quantum leap in thinking and give a business a lot of information and motivation to improve performance:

- How good are we?
- How fast are they?
- How many engineers do they have?
- How quickly do they changeover?
- What is their wastage?
- How many customer complaints do they get?
- What is there energy consumption per tonne of product?
- What is the best?

Information in a benchmarking exercise is often presented as a league table, with the best performing at the top. By carrying out regular benchmarking exercises, you can see if your factory is moving up the table. You have to remember that because business, and especially the food business, is very competitive, to move up the table you not only have to improve, you have to improve faster than your competitors. Getting better does not necessarily mean you will go up the table; you have to be the best at getting better too!

Cost Reductions – in the competitive world of food manufacture, the retailers want year-on-year cost reductions for your products to allow them to attract consumers to their stores. Benchmarking can help a food business to improve performance so that it can offer a price reduction but still maintain the profit level that it needs.

Sources of benchmarking information

It is relatively easy for benchmarking comparisons to be made inside a business to assist in raising the bar for the people working there. Comparisons from machine to machine, department to department, shift to shift are simple to organise and are probably done as a matter of routine. The best shift for productivity is Red shift, the Blue shift runs with less waste, and White shift gets better quality scores.

With larger companies it is possible to compare performance with other factories, other regions and other divisions. The Bristol factory is best for productivity, Germany has the lowest waste, and chilled division has the best absenteeism rate.

The more difficult area of benchmarking is that of comparing your performance outside of your company. Getting information is difficult but not impossible. Benchmarking groups are often set up within an industry that will periodically ask its members to submit data for comparison. Confidential reports are produced, often with league tables, to allow people to see what good looks like.

There are other sources of useful information that include newspapers and trade journals, which often carry stories about your competitors. By a bit of detective work, it is possible to piece some information together. One story might carry a figure for the number of employees; another story might mention the energy consumption or effluent volumes. As soon as there is information on the quantity of products made, it would be possible to get a number for productivity, energy per tonne, water used per tonne. These numbers can then be used to benchmark against your own performance and maybe highlight opportunities to improve.

Information about other companies can come from many sources. Keeping an eye on job advertisements will maybe give some hints. A limited company has to produce an annual report. The report can contain much data that will be useful in a benchmarking exercise for your business. Your customers and suppliers can provide information. Auditors and equipment suppliers can also be a source of information. Finally, the food industry is close knit and people move around the industry from company to company. New employees, recruited from other companies, can be a useful source of benchmarking data.

The key thing to remember is that knowing you are fourth in the productivity league table is only the start. You need to investigate why that might be; spot opportunities to improve, then install and maintain those improvements. Then hope that you have carried that improvement out fast enough to be third next time and not fifth.

World Class performance

The overall aim of Lean Manufacturing techniques is to improve performance to a point where your business has achieved a position in the industry where that performance has been achieved by a process of continuous improvements rather than large-scale re-investments. The ultimate aim is to achieve the status of World Class. World Class manufacturers are those that demonstrate industry best practice. To achieve this, companies should attempt to be best in the field at each of the following:

- Quality
- Price
- Delivery speed
- Delivery reliability
- Flexibility, and
- Innovation.

Food factories should therefore aim to maximise performance in these areas in order to maximise competitiveness.

It is often said that the top 5% of companies can consider themselves world class, but you will notice that nowhere in the six areas above are mentioned the words profit and cash flow. The assumption is that if your company is world class, then business will flow to your company, you are one of the cheapest with the best delivery, and the best quality after all.

In the food manufacturing sector, with all of its pressures, achieving the status of world class is a real achievement that has been attained by very few. There are many companies on the journey to world class manufacturing status, so the race to improve is on. It is a journey that all can take. The route to world class is often low cost or no cost. It does require a rigid discipline to make sure that improvements in performance do not get lost on the choppy seas of the food manufacturing industry.

Summary – How good are we?

In this final chapter, the question was asked that occurs to all people who are trying to improve: "How good are we?" The methods that are commonly used for benchmarking factories were examined, so that it is possible to see where you are in the league table of factories and also that you can see which aspects of your performance are letting you down in your work to get to the top. Is it your control of waste? And if it is, which form, out of the seven, is hurting your performance most? Is it the speed of your changeovers that is causing a larger than average availability loss? Are you performing badly because you have large quantities of work in progress, or is it the level of staff motivation that is causing you not to be world class.

Lean Manufacturing is all about making continuous improvements in your business to drive up performance and drive down cost. Setting targets is a big part of the technique that encourages people to come up with no cost and low cost solutions to your issues.

Remember, standing still is never an option. You have to get better just to stand still. A correctly installed system of Lean Manufacturing can help you improve faster than your competitors and it is that speed of improvement that will see your business grow.

Appendix 1

Management style feedback sheet

Answer the following questions by indicating your feelings on the scales provided. The questionnaire should be completed quickly and your first thought recorded as your response. The questions are trying to establish how you feel at this moment.

Your answers are confidential and will be used to look at ways in which the management of the factory can be improved:

1 In matters where I feel my opinions should be sought, I believe that my opinions are valued by my supervisor.

   Always   Nearly Always   Almost never   Never

2 I believe that I am treated fairly by my supervisor.

   Always   Nearly Always   Almost never   Never

3 I believe that my workmates are treated fairly by my supervisor.

   Always   Nearly Always   Almost never   Never

4 When I have a suggestion of how things could be improved, I am happy to talk with my supervisor about it.

   Always   Nearly Always   Almost never   Never

5 I feel that my suggestions are taken seriously by my supervisor.

   Always   Nearly Always   Almost never   Never

6 I feel that my supervisor sets a good example to us.

   Always   Nearly Always   Almost never   Never

Handbook of Lean Manufacturing in the Food Industry, First Edition. Michael Dudbridge.
© 2011 Blackwell Publishing Ltd. Published 2011 by Blackwell Publishing Ltd.

7 I believe that my supervisor sets high standards for us to hit and lets us know if the expected standards are not met.

   Always   Nearly Always   Almost never   Never

8 I enjoy working with my supervisor.

   Always   Nearly Always   Almost never   Never

9 I believe that my supervisor is fair in the allocation of tasks.

   Always   Nearly Always   Almost never   Never

10 My supervisor gives clear instructions.

   Always   Nearly Always   Almost never   Never

Thanks for your time in completing the questionnaire.

Appendix 2

The calculation of some typical KPIs used in the Food Manufacturing Industry

KPIs are indicators of performance that are used in the management of production operations.

KPIs can be calculated in many different ways - the important thing is the consistancy of the rules that are used to calculate them.

Production Efficiency. That is how much did we produce compared to how much we could have produced in the same time had everything run perfectly.

Number of packs or tonnes or batches or sausages produced divided by the theoretical maximum that could have been achieved in the same time.

So its possible to calculate the production efficiency every day, every shift or every hour depending on your need.

Labour Productivity. That is the output per man-hour.

The total output in packs or tonnes or batches or sausages divided by the total number of labour hours used.

The total labour hours used number is the head count multiplied by the period being studied.

So if you have a crew of 10 people for an 8 hour shift that is 80 man hours.

Material Yield. How much did we use compared to how much we should have used to make the product that we made.

Fairly simple this one – calculate the quantity of cheese you should have used to make the pizzas and divide that number by the actual amount of cheese used.

So, if you should have used 100Kg and you actually use 105Kg that's a 95.2% yield.

Product Quality. How much did we make right first time compared to our total output.

This is a measure of the "right first time" on the production line.

It's the total number of production minus the products reworked and then divided by the total number of production.

Availability – The percentage of time the machine could run compared to the amount of time we wanted the machine to run.

This number is often calculated by measure the length of time that the machine was not available and deducting that number from the total time of the shift.

That will give you the UP time for the machine. If the UP time is divided by the total time of the shift that will give a number for the availability percentage.

Appendix 3

The XYZ Food Factory Daily Weekly Operating Report

Week Comm

KPI		Target	Sun	Mon	Tue	Wed	Thu	Fri	Sat	Total
OEE	Red									
	Blue									
	White									
Labour Efficiency	Red									
	Blue									
	White									
Waste	Red									
	Blue									
	White									
Quality	Red									
	Blue									
	White									
Accidents	Red									
	Blue									
	White									
Absenteeism	Red									
	Blue									
	White									
Energy consumption per tonne	Red									
	Blue									
	White									

Index

Handbook of Lean Manufacturing in the Food Industry, First Edition. Michael Dudbridge.
© 2011 Blackwell Publishing Ltd. Published 2011 by Blackwell Publishing Ltd.

Printed and bound by CPI Group (UK) Ltd, Croydon, CR0 4YY

23/04/2025

14660953-0001